With best wishes!

Dan Mackie

H. G. Quesneer

ENGINEERING
MANAGEMENT
OF CAPITAL
PROJECTS
A PRACTICAL GUIDE

DAN MACKIE, P. Eng.

McGRAW-HILL RYERSON LIMITED
TORONTO MONTREAL NEW YORK AUCKLAND BOGOTÁ
CAIRO GUATEMALA HAMBURG JOHANNESBURG LISBON LONDON
MADRID MEXICO NEW DELHI PANAMA PARIS SAN JUAN
SÃO PAULO SINGAPORE SYDNEY TOKYO

ENGINEERING MANAGEMENT OF CAPITAL PROJECTS

ISBN: 0-07-548846-9

2 3 4 5 6 7 8 9 THB 3 2 1 0 9 8 7 6

Care has been taken to trace ownership of copyright material contained in this text. The publishers will gladly take any information that will enable them to rectify any reference or credit in subsequent editions.

The personal pronoun "he" has been used in most cases throughout this book for the purpose of simplicity only. Neither the author nor McGraw-Hill Ryerson Limited intend to imply that the professional roles described herein are exclusive to one gender.

Canadian Cataloguing in Publication Data

Mackie, Daniel
    Engineering Management of Capital Projects

Includes index.
ISBN 0-07-548846-9

1. Industrial project management   2. Engineering – Management   I. Title.

TA190.M32 1984   658.4'04   C84-099437-0

To Lori, who believed in this book
and, from the beginning,
believed in its author

ACKNOWLEDGEMENTS

Instrumental in laying the groundwork for this book over the past twenty years were my many mentors, not the least of whom were: Earl Jelter, Willy Weydemuller, Earl Johnson, John Watson, and Boyd Henderson. Particular thanks go to : Boyd Henderson and Ed Amos, who reviewed the manuscript and offered many useful suggestions; Rita Papantoniou, who toiled cheerfully at typing from my scrawl; and finally, my editor, Peter Matthews, who lent unflagging support to the project and proved once and for all that talented editors still exist.

Dan Mackie, P.Eng.
February, 1984

FOREWORD

There is an old saying about the two types of project managers: if one is calm, collected and confident that he is on top of his project, while the other is scurrying frantically and tearing his hair out, you can be sure that no one has told the first man what is going on.

There is enough truth in this comparison to make clear the need for everyone involved in running projects to understand the underlying principles of project management and to be reminded of them at frequent intervals.

The construction industry, which creates wealth and improves industrial productivity, is one of the major driving forces in any economy. This is particularly true in Canada. The direct employment it gives in the engineering office, in the manufacturer's plant, and on the site itself probably makes it the largest single industry in the world economy. In addition, it provides new plants and facilities for the efficient manufacture of modern products, for transporting them more economically, or even, as in the case of the garden shed in Chapter 1 of this book, for storing them better. Unfortunately, the construction industry in North America is not in good health. This is due not only to the downturn in the economy: over the past decade the industry has experienced a sobering drop in site productivity at a time when the introduction of new equipment and techniques should have heralded a dramatic improvement.

In the United States the Business Roundtable, a forum which brings together the chief executives of many of the largest private companies in the country, has just completed a four-year, in-depth study of the construction industry. It concluded that poor project management, in conjunction with outmoded labor practices and unnecessary government restrictions, has resulted in a 20 percent increase in real field construction costs over the last ten years. In other words, twenty cents out of every dollar spent on the site for new plant construction has been wasted. This loss has had a direct impact on our domestic costs and on our competitiveness in the foreign marketplace. And perhaps most serious of all, it restricts the amount of capital available for other investment. This decline is of vital concern to all of us if we are to maintain our high living standards. It underlines the serious challenge facing project managers in today's environment.

A simple but accurate definition of project management is that it establishes what is to be done and how it is to be measured — no more and no less. In recent years, the development of sophisticated and often computerized project control techniques has focused attention on project management as a separate discipline. This has encouraged people to stand back and analyze actions which previously had been almost instinctive. It has also given rise, unfortunately, to a plethora of publications which are minutely researched, erudite in the extreme, largely incomprehensible, and, ultimately, useless. It often seems that, in striving to find the best way to do the job, we are forgetting what the job is — to build a good plant, on time, and within budget.

It is necessary to go back to fundamentals and to develop a basic understanding of how to manage a project. This book will give the young manager this understanding and will remind the experienced manager of the underlying principles involved in project management.

W.B. Henderson, P.Eng.
February, 1984

PREFACE

Some time ago, when I was asked to present a seminar on project management to a mining company, I approached the task with some trepidation. I knew the company well, and that's just what bothered me. They already had a fairly good project team on staff which was performing not too badly. What, I wondered, could I tell them, short of nitpicking at their operation?

They had some weaknesses, to be sure, but I decided to avoid any head-on confrontations. Instead, it seemed to make sense to give a general dissertation on project management, using their own people in the presentations and hoping that this would lead them to discover some truths about themselves. I have always found this technique to be the most effective and to give the most long-lasting results. It is an easier pill to swallow than criticism coming from someone else.

Out of a group of forty attending the seminar, three were project managers and the rest were engineers, estimators, schedulers, cost controllers, administrators, and construction superintendents. Guests included a couple of senior people, the president, and the chief executive officer. Only two from the specialty groups were slated to become project managers at some time in the future. According to the president of the company, all were satisfactory-to-excellent in their fields of specialization.

It had been the president's objective to strengthen his projects group so that cost and schedule overruns would virtually disappear, a tall order in those inflationary and unpredictable times. But there were a large number of capital projects planned for the next five years, and this added to his concern. With a good team in place, it made me wonder what key element would stimulate some sort of overall improvement. An infinitesimal improvement — maybe. Far reaching? — hardly.

The presentation given was a standard outline of procedures in each specialized area of project management. There was a description of how things were being done in the company versus how they *should* be done. In many cases, the two were the same, give-or-take a few details.

Despite strong leadership in their projects, the company's project managers were found to lack an understanding of some basic principles. Trend forecasting was their main weakness — the area that is most often misunderstood by both project managers and company officers. Overall, seeing a comprehensive package on project management and the debates resulting from the presentation sensitized the project managers to the weaknesses of individual departments and uncovered some weaknesses in task force members.

Specialty leaders, on the other hand, were made acutely aware of their lack of understanding of the project management process and of how their own work was coordinated with other disciplines. Many expressed surprise and delight ("So *that's* how it works!") and admitted that they had been too busy doing their own jobs in the past to consider what the rest of the task force was doing.

Even more surprising was the positive response received from the senior officers of the company. It seems that they, too, benefited from learning how project management works.

Clearly, the exercise revealed a definite need for an understanding of project management, not just by the project managers and participants, but by the *whole organization*. This notion was reinforced by requests from other departments — financial, maintenance, and research — for another seminar. At the senior officer level, it was felt that the senior corporate officers needed to be reeducated so that they could better understand the forces that generate requests for large capital expenditures.

Two follow-up seminars were conducted: one for the remaining departments and one for senior management. A project management procedures manual was prepared, using input from the specialty groups, and corporate policy regarding capital spending was included in it, bridging a wide communications gap that had hitherto gone unrecognized. The manual has since become a valuable handbook for all levels of the organization.

Some time later, I received a telephone call from a vice president of an engineering company which had been engaged to do work for the mining company that I have just described.

"Operating companies are usually their own worst enemies when it comes to dealing with projects," he said. "Usually, we have to do considerable management of our clients. It sure is refreshing to work with one who really understands project management. Thanks."

After a further exchange of pleasantries, he added. "You know, you really should write a book. The industry needs it."

CONTENTS

CHAPTER

1

SETTING OFF IN THE RIGHT DIRECTION

DEFINITIONS AND OBJECTIVES

To begin with, let us define project management as both the art
and science of spending capital resources to realize a useful physical structure. To say that it is an art as well as a science is to emphasize that a great deal of intuition and good judgement is essential
to achieving success in project management. To say that a useful
physical structure must result is to suggest that the project management process can cover a gamut of projects and, further, that
the size of those projects is irrelevant. Indeed, the principles outlined in this book are applicable to industrial and commercial projects ranging in capital cost from a few hundred thousand dollars
to megaprojects worth many millions. For the sake of clarity,
however, very simple examples will be used to illustrate the way
things work.

A project manager usually heads up a team that performs capital works. The overall project is often referred to as EPC —

Engineer-Procure-Construct — although confusion of definitions runs rampant in industry. Project managers are referred to by electronic data processing firms as those responsible for managing or setting up computer systems. Some engineering consulting companies dilute the title of project manager by making anyone who has control of or responsibility for a piece of work a project manager, even though the work may be part of a larger project. Other companies refer to the project manager as the project engineer, while still others have project engineers who work for project managers. In this book, a project manager is someone who has total responsibility for EPC from the beginning of the project to start up of the plant, mine, or refinery or to the opening of the building, dam, highway, or park.

A successful project can be defined as one that is:

- On time
- On budget
- Meets design criteria

The project manager is not the only one who needs to understand the process of project management. Indeed, he can be rendered quite impotent by trying to manage in an atmosphere of ignorance. If he is trying to play a game of stud poker and everyone else at the table thinks it is gin, then he has lost before the first card is dealt. Too often, the success of the project is thought to rest entirely on the shoulders of the project manager, and after all, what does responsibility matter in big business? It is that kind of thinking that results in projects that are estimated at $46 million and come in two years late at $90 million. You may ask: So what? Doesn't everyone get paid? Doesn't the owner still make money?

Well, if that company makes automobile tires, you, the public, will have to pay more for them. If they make cheese, then the price of cheese will go up. And if they make aluminum, the cost of airplanes and homes will rise. In other words, a botched project hurts everyone.

Finally, it is assumed in this book that the project manager is on the owner's side of the project, that management will be directly under his control, whether or not a consulting company is needed to assist. The methods and systems described here can be used by an owner to run his own projects directly, or they may relate to projects run by a consultant. Either way, it behooves the owner to understand how it all works.

RECENT HISTORY OF PROJECTS

Even before the time of the pyramids, man toiled for man. Project management as a modern skill, however, did not come into its own until World War II, when it was born of necessity. It played a part in the Allied war effort, a classic example of what people can do with their backs against the wall and a will to pull together.

As the postwar economy floundered, a legacy of skilled project management teams remained in the huge consulting and construction companies — the Bechtels, Kaisers, CF Brauns, Brown and Roots, Kelloggs, and Foster Wheelers, to name a few — so that by the time the 1960s came along a formidable roster of talent existed to nurture a new surge in economic wealth. Viet Nam needed to be fuelled by steel and lead and copper and oil. Then there was the space program, the man on the moon, and finally, the shuttle.

Meanwhile, the size of projects began to grow. Economies of scale were being exercised to the ultimate. From the St. Lawrence Seaway in the late 1950s to the iron ranges in Michigan, Wisconsin, and Labrador, to the James Bay Hydro Project and the tar sands of Alberta, projects swelled in size and complexity and daring. The larger the project, the greater the risk; so too, the greater the need for project management skills.

If a large project is a very serious business, a megaproject is many times more so. Consequently, the recent history of megaprojects has been plagued by a fear of failure which slowed down the decision-making process. But at the same time, this fear increased the apparent need for back-up skills. Projects became overmanned, bureaucratic, and artificially complicated.

The apparent need for tighter project control on all sizes of projects began to show itself with the first signs of rampant inflation in the early 1970s. When the oil cartel in the Middle East created an energy crisis, control became a very tricky proposition indeed. The embarrassment of going back to boards of directors for more money became commonplace. While the numbers and sizes of projects continued to escalate, so did the overruns of cost and schedule. It became almost fashionable to talk in terms of minimizing cost overruns instead of eliminating them.

Naturally, the more outraged businessmen became, the more heads began to roll and the more cautious project management became. To use the vernacular, a "protect-your-ass" syndrome developed.

With two forces active in project management — the size of virtually out-of-control megaprojects on one side and rampant inflation on the other — the whole process bogged down in the resulting bureaucratic mess. Clearly, there is a desperate need for retrenchment and optimization, a need for a rebuilding process that a number of industries have recognized and attempted to meet with sober determination.

PARABLES ON THE ART OF PROJECT MANAGEMENT — GAUGING HOW FAR TO GO

Harry Bischophylus and Ron Goodfellow

Harry and Ron were next-door neighbors. Harry's lawn had been taken over by crabgrass and dandelions. He had planted some flowers, but they died. Harry had no place left in his garage for his rakes and shovels and hoes, so he stood them in a corner of his yard.

Ron Goodfellow's grass was as smooth as a putting green. He could be seen cutting it twice a week, whistling as he pushed his mower around his yard. The Goodfellows lived in an immaculate house. Harry's wife, Anita, was fond of pointing this out to Harry.

Secretly, Harry Bischophylus hated Ron Goodfellow. Whenever Ron did something, Anita pressured Harry into trying to do the same. When the Goodfellows got a new station wagon, Harry had to buy one too. When Ron got a new wheelbarrow, so did Harry. When the roofers arrived to replace the Goodfellows' shingles, they could be sure that their next job would be at the Bischophylus household.

One day, while chatting over the fence, Harry learned that Ron was planning to build a new shed in the backyard. Knowing that when Anita saw it he would have to build one too, Harry resolved that, for once, he was going to beat Goodfellow at his own game. "I'm going to construct a shed too," he thought to himself. "But I'm going to build it quicker. Why, I'll have it up while Goodfellow is still thinking about it. Anita will be thrilled.''

So Harry went to the hardware store and bought some reinforcing wire and some boards for forms. Then he began digging.

"What are you doing?" asked Anita, surveying the excavation.

"It's going to be a surprise, my pet."

When he had finished digging and putting up the forms, he called the cement company. At first they said they couldn't deliver for a week, but when Harry said that it was urgent, they said that it could be done but that they wouldn't deliver less than five yards of concrete.

Later that afternoon, the ready-mix truck arrived. It couldn't get to the backyard, so Harry had to run concrete back by wheelbarrow. It was a hot day, and the work was very hard, but at last Harry managed to pour the concrete into the forms that he had built. "What," asked the truck driver, "do you intend to dc with the rest of this concrete?"

"What do you mean?"

"There's still two yards in the truck."

"I don't need it."

"If I don't dump it quick, it will harden," growled the driver. "You take it now, or I'm going to dump it in your driveway!"

So poor Harry wound up with a slab for his shed plus a rather irregular patio in his garden. He did not feel too badly about it, though, because it covered his dead shrubs. Besides, Ron Goodfellow had not even started his shed. In fact, it occurred to Harry that he had not seen Ron all day.

On the following morning, Harry went to the builders' supply and picked up the 2 x 4s he thought he would need, a pound of three-inch nails, and some sheets of particle board. He cinch-anchored some 2 x 4s onto the concrete pad in the shape of the shed, then began cutting and nailing studs in place. Then he ran out of nails. Just to be sure he would not run out again, he bought twenty pounds more. By the end of the day, the shed was beginning to take shape. Harry felt tired, but he had yet to see Ron Goodfellow working on his shed. He wondered if Ron could use the eighteen pounds of three-inch nails he had left over.

All that week, Harry spent his spare time cutting the particle board to shape and nailing it onto the sides and roof of the shed. On Saturday morning, as Harry began nailing the shingles onto the roof, Ron Goodfellow appeared at the side of his house. He removed some pins from his fence, swung the fence aside, and backed his station wagon into the corner of his yard. Harry watched from the roof of his shed as Ron unloaded some precast concrete patio slabs and bags of crushed stone. Using a rake and shovel, he levelled an area, covered it with a uniform layer of crushed stone, then fitted the concrete slabs in place. "Might want to move this foundation some day," he said to Harry, who had just hit his thumb with a hammer. In an hour, Ron had completed his foundation.

Harry then watched Ron carrying the prefabricated side of a shed from his garage. It had a temporary stand nailed to it so that he could place it upright on the foundation. Within half an hour he had all four sides up and had begun adding the prefabricated trusses that he had built during the week in his garage.

Naturally, Harry felt a little nonplussed, but all he had left to do to finish his shed was to add the siding, window- and door-framing, and the windows and doors.

While Ron was nailing his precut sides in place and starting the roof shingles, Harry went to the builders' supply for materials — window- and door-framing, siding, windows, and doors. Although Ron Goodfellow's shed now looked like a shed, he still had a long way to go, in Harry's opinion, so Harry did not feel too panicked as he began cutting pieces for the window frames. At the end of the day, Harry had finished all the frames, and Ron had only just completed his roofing. Harry was sure he would complete his project ahead of Ron!

The next morning, Harry installed the windows. He had a bit of a problem fitting the door, but he managed to get it in place and working. Just as he began nailing on the aluminum siding, Ron Goodfellow emerged from his garage carrying a prefabricated window-frame with the window already in place. Harry was only halfway through attaching his siding by the time Ron had all the doors and windows installed. Shortly after lunch, Harry sat on his back step cursing silently to himself. He had run out of aluminum siding. It was Sunday — the builders' supply was closed — and with considerable gnashing of teeth, Harry watched helplessly as Ron Goodfellow added the last bit of siding to his shed!

Just then, Anita appeared at the back door. "It looks kind of nice, dear," she said. "But it's not nearly as nice as the Goodfellows', and it's in the wrong corner of the yard!"

Moral: Even a small, one-man project can benefit from the project management process. You can bet that Harry's shed cost a good deal more than Ron's. Chances are, Ron did a design drawing followed by a material take off and proper sourcing of his material. He worked out a plan so that he had a clear idea of where he was going, and then he stuck to it. In its simplest form, this is the project management process.

True, both men achieved results. But Harry simply put his head down and charged full speed into the project. Ron, on the other hand, took a methodical approach, planned everything ahead of time, and attained his goal in a controlled, confident fashion. Similarly, those who work on industrial and commercial capital projects can get results without skillful project management, and with a little luck and perseverance they can actually be successful and satisfy the investors. The trouble with this approach is that you are never really sure of what the results will be until it is too late.

Harry could only have been accused of blowing his schedule, since his objective was to have his shed completed ahead of Ron. Just wait till he hears from Anita when the bills come in, however. And when do you think she's going to ask him to move the shed? We have to wonder, too, about his choice of colors. Will he have to replace the siding?

George Gillycuddle

If you think Harry had problems, you should talk to George Gillycuddle, who lives six blocks over. George is thirty-seven years old. He lives in a neat bungalow with his mother. George is very good with his hands, but when he built a laundry room, it cost a little more than he had planned, and the shelves never sat quite right. When the washing machine went into its spin-dry mode it set up sympathetic vibrations in the floor joists, and the whole house shook. George's mother never mentioned this to him, but he could feel her eyes on him every time the house began shaking.

One day George's mother mentioned that they needed a shed in the backyard and that they should buy one of those ready-built models from a department store and have it put up by a local handyman.

George took no pains to hide his deep sense of rejection, and, recognizing her error, George's mother suggested that perhaps he would like to build a shed from scratch. George immediately cheered up and began making plans to build a shed. He was determined to produce the finest shed in town. Mrs. Gillycuddle was pleased to see her son so happy, but she was convinced that he would screw it up as he had with the laundry room.

George thought he could build a shed for about $1,000 and that it would take a month to construct. Afraid that he might exceed these targets, however, he told his mother that the cost would be "in excess of $3,000" and that the project might take as long as six months due to the possibility of a suppliers' strike. Confident that he had given himself plenty of leeway, George went to work.

George wanted to be doubly sure of placing the shed in the right spot in the yard, so he talked it over with his mother. When she couldn't decide, George drew a plot plan of the backyard and took some photographs, marking the direction in which the pictures had been taken on the plot plan. Together they pored over these photos and drawings, and, when they were sure they had it right, they went to the next stage, which was choosing a design for the shed. George drew up about a dozen designs and made paste-ups, which they stuck on the photographs. After a week of consideration, a selection was made and he began drawing up the detailed design.

Once George had completed the design, he wrote lengthy specifications for the supplies and then went out for competitive bids. He rejected seven plywood bids out of hand because they had not filled out all the blanks in the bid form that he had prepared, and he had to go back to the rest for clarification on the type of glue used in the manufacture of the plywood. The hardware store where he wanted to buy the nails also refused to sell by the nail instead of by the kilogram, but in the end, after he had spoken to the manager, they counted out 394 nails for him — the exact number he had calculated he would need.

Six months later, George Gillycuddle had progressed only as far as the foundation of his shed, and then he had to quit the project for the winter because of snow. He didn't mind, though, because it would give him time to update his critical path schedule so as to be ready for spring.

Moral: Any project, large or small, can be overmanaged. And, strange as it may seem, mistakes are just as apt to happen in an overmanaged project as they are in an undermanaged one.

The story of George Gillycuddle may sound exaggerated. It may even seem frivolous, given the seriousness of project management in the real world, where millions of dollars are at stake. Yet some of the examples used in this story are drawn from the case of a real mining project. The project manager was dreadfully afraid of making a mistake. His staff quickly became weary of his constant quest for perfection and began to view most of his demands as redundant. Consequently, what seemed like a perfectly planned design failed. Not only did it overrun its budget and schedule, but it did not meet the design criteria.

The managing of capital projects is largely dependent on good judgement and a feel for optimizing planning and control. Under-controlling projects produces slapdash work. Overcontrolling usually results in irresponsibility among subordinates and encourages them to figure out ways to beat the system. You should remember that the object is to get the job done in a reasonable time at reasonable cost, producing a project that is acceptable to its user.

CHAPTER

2

THE SCIENCE OF PROJECT MANAGEMENT

FIVE BASIC ELEMENTS OF PROJECT MANAGEMENT

There are five basic requirements for conducting a successful project. In order of importance, they are as follows:

1. Choosing the right people
2. Choosing the right people
3. Choosing the right people
4. Setting up the right organization
5. Using the right systems

Only three requirements, you say? Perhaps, but the importance of selecting the right people cannot be emphasized enough. You can run a project with a poor organization and inferior systems and have reasonable success, but good organization and superb systems can never substitute for having good people in place.

Selecting and Using the Right People

At every football, baseball, or hockey game where there are, say, 50,000 spectators, probably 20,000 of them are ardent fans, and the majority of these are self-avowed experts on the game at hand. Yet, if any one of them were to actually be parachuted into the coach's position, the odds are his team would not win. So it is with project managers.

For some reason or another, everyone in industry thinks he is a project manager, and, more often than not, industry helps him along in that belief!

Several years ago, a major metallurgical plant was in the offing in northern Canada. A plum for the consulting industry, the package consisted of engineering, procurement, and construction to the successful bidder. As it happened, a large consulting company had been trying aggressively to establish itself in that area. Since it was an economic boom time, the odds seemed pretty good for that company. Rumor had it that some political string-pulling was taking place and economic concessions were being made. The consultant, it seemed, had the project in the bag.

A young project engineer, anxious to make his mark and excited about the project, could not contain his curiosity about his probable role in this multimillion-dollar project, so he begged one of the secretaries to let him see the proposal. He knew that he would be important to the project because his area of expertise was in the design of metallurgical plants. It would be, he reckoned, a chance to show his worth.

This is what he found: As a concession to the owner, the consulting company had asked a vice president to step down to project manager, reasoning that a project as large and complex as this one required a senior executive to run the show. Now this vice president had worked his way up the corporate ladder as project manager of an aircraft hangar construction project and in the design and erection of a cookie factory, both of which had been successes in the eyes of the company. This metallurgical plant would be his first and, he reckoned, another feather in his homburg.

Our intrepid project engineer was promoted to area engineer, along with several others, each of whom got an area of the plant complex to design. The area assigned to him was utilities — water supply, power supply, and sewage disposal — an area that he thought he could handle but which was definitely outside his area of expertise. He was quick to note that the other area engineers, like the project manager, had never seen the inside of a metallurgical plant. He refused to participate in the project, asking for a transfer to another city.

Why? As an aspiring project manager, the project engineer knew that there was little likelihood that the technical control of the project could be monitored and maintained without a great deal of input from the owner. Since the proposal showed no such formal arrangement for this kind of input, the engineer reasoned that failure seemed inevitable. He left the project before it started.

Eighteen months later, the consulting company felt the toe of the owner's boot and had to leave the property in disgrace. Clearly, the wrong people in the wrong slots did this project in.

CHOOSING A PROJECT MANAGER   There is a belief that is common to project managers that being a professional manager means having the ability to handle any project, be it an oil refinery, a water treatment plant, or a condom factory. It is only a question of hiring experts to handle the technical end of the job. The rest is simple logic, combined with an ability to control the budget and the schedule.

At the other end of the spectrum, there is a school of thought that says there is absolutely no way a person can build, say, an iron ore concentrator without an intimate knowledge of iron ore concentrators, and there is no amount of project management training or experience that will compensate for lack of such knowledge.

Pundits will argue through the night one way or another, and, as with religion and politics, they will rarely win over the other side. Each side can point to numerous completed projects that have been successful with one type of project manager or another in harness.

The fact of the matter is that the very nature of his work demands that a project manager be a generalist. Usually he can stumble along with just technical experience and a modicum of project experience or vice versa. His chances of success are greater, however, if he has both.

In choosing a project manager, therefore, one should look for the following:

– training *and* track record in project management
– background in the type of project or in a related industry
To insist on your candidate having those qualifications is to en-sure a successful project. Any other choice is a gamble.

CHOOSING OTHER PERSONNEL Obviously, the project manager's prime task is to come up with a good team. If he is to be a manager in the true sense of the word, then he must realize that it is up to him to fill in the gaps where there are weaknesses in the team. Ideally, he will see that the following essential areas are covered by personnel suited to the discipline:
– Administration
– Engineering
– Process
– Design
– Estimating
– Planning and scheduling
– Cost control
– Commissioning
– Construction

On smaller projects some tasks will be grouped, and on larger projects some will be expanded. For example, engineering, pro-cess, and design could be controlled by one person.

It goes without saying that the goal of the project manager is to place a top-notch person in each slot and, even more important, to assemble a team that works well together, possibly one that has worked together before. Ideal situations are rare, however, and there is usually a weak leader in one or two of the disciplines. In that situation, the project manager needs to support the disci-pline. He needs to know his own strengths and weaknesses, how-ever, recognizing that a weak discipline combined with a project manager who is also weak in that discipline will spell disaster. That, in turn, underlines the need for a project manager who has experience in a related industry. Suppose he winds up with a weak design team? How will he know where their weaknesses are if he doesn't have pertinent experience? How can he judge if the de-sign is a Cadillac when a Chevy will do?

CHEMISTRY A chemical plant planned for southern California was headed up by an engineer who had worked in operations in his youth and then been hired by a major consulting company that specialized in the design and construction of chemical plants.

As a junior in the consulting company, he underwent intensive training, spending several months in project management systems: estimating, planning and scheduling, and cost control. Then he worked as design manager in engineering, rising very rapidly to project manager. He had made a success of two projects; this would be his third. Certainly he was considered the ideal man for the job.

Now, the economy being what it was, our project manager found it very easy to staff the job. In fact, he was so pleased with his incredible luck at putting top quality personnel in all categories that he announced to his wife that she would see much more of him than she had during past projects. "Gone will be the late night and the weekend sessions at the office to correct mistakes and redirect efforts of the weak links in the chain. This job is going to be a piece of cake. The project will run itself."

In fact, the project started out quite well, and our manager had no qualms about putting his second-in-command, the design manager, in charge while he took a short vacation with his wife.

When he returned a week later, however, the construction manager was waiting for him in his office. After pleasant formalities were exchanged — the project manager felt good after his rest — the construction manager suddenly leaned forward in his chair and said, "If you don't take that sonovabitch off the job, I'm going to quit!"

A peaceful Monday morning instantly became chaos. The design manager and the construction manager, it seemed, had had a dispute early in the week over proposed construction methods. According to the construction manager, the design manager had issued field instructions without consulting him. The project manager said that he would take it up with the design manager right away.

It seemed to the project manager that his design man was a reasonable person. When he confronted him with the problem, however, the design manager accused the construction manager of taking short cuts in the field that would endanger the project.

On examining the evidence, our man came to the conclusion that both managers were right about their proposed methods. It was strictly a case of "there's more than one way to skin a cat." Probably, he reasoned, if the design manager had approached the construction manager first, before issuing instructions to the field, the problem could have been resolved. Clearly, he needed to make an arbitrary decision at the risk of offending one or the other, but he felt that both men were reasonable. They just needed to realize his position and allow the project to carry on.

When he brought the two men together, he explained that he felt obliged to go with the construction manager's decision for no other reason than that it would look good in the field to have the leader's point of view supported. He felt, however, that they should be capable of reaching an agreement without an arbitrator, and he chastised them for not resolving the problem in a calm, logical manner.

Feeling good about the outcome of the meeting and comforted by the fact that the two men shook hands before they left his office, the project manager went about getting back into his office routine. The next morning, however, his secretary told him of reports of the two men shouting at each other across the office.

As the project progressed, the schedule began to slip. This puzzled the project manager because he knew that the planner and scheduler was a good man who rarely missed predicting accurate interim or completion dates. After a couple of weeks, he discovered that the design manager and the construction manager were stonewalling each other, each digging in his heels on minor points. Nothing got accepted in the field without extensive modification requests, and any changes proposed by the construction team met with resistance in the design office. It was too late to replace either of the managers. The project manager had no choice but to personally take control of the work, pushing through decisions despite the objections of one or both of his leaders.

"How come we never see you on weekends?" his wife asked.

If the chemistry doesn't work, then good men will become stifled. It is important, therefore, to try to put together personnel who are competent *and* compatible. A team that has worked together before is usually a good bet. If that is not possible, the project manager should watch for signs, however subtle, of personality clashes. Sometimes it pays to intentionally stimulate controversy in order to ferret out differences early in the project, before things begin to deteriorate.

Selecting the right people goes beyond the individual in his area of expertise. The chemistry within the team has to be right too.

Setting Up the Right Organization

In the old days, one seldom saw The Big Man, but his presence could be felt. When he hollered, the whole office jumped. The boss was the boss was the boss, and usually his awesome authority made itself felt in the form of a reprimand. One never questioned authority.

That type of organization spawned the hierarchy, a pyramidal structure with the president (or king) at the top, taking responsibility for and having authority over the whole enterprise. Those who were insecure trusted no one; the more mature chose lieutenants and sergeants to run things for them. Occasionally, the hierarchy failed in its duties. The less secure people at the top who remained in power sometimes created a matrix organization to make sure that the lieutenants and sergeants were doing their jobs.

In project management, Figure 2.1 is a typical simplified hierarchy; Figure 2.2 is a typical matrix. In a hierarchy, orders flow from the top to the bottom; results are reported from the bottom to the top. In a matrix organization, the same is true except that there are discipline chiefs who can issue orders parallel to the project manager's.

Both organizations have become more democratic over the years. Although the same power structures exist, there is now a tendency to get more feedback from the floor. In the extreme cases, such as in Japanese organizations, there is direct participation by "the floor" in management decisions. This appears to be more satisfying to the lower echelons of the structures and is reputed to be instrumental in creating high productivity in Japan.

There is a distinct tendency for hierarchies to become matrices, particularly in high-risk industries, such as the nuclear or aerospace industries. And the same trend is evident in government organizations. Decisions tend to be made by committee. This dilutes responsibility and it becomes increasingly difficult to single out individuals for blame when things go wrong. Ironically, the more an organization is matrixed, the more things seem to go wrong — in governments, for example. What more needs to be said?

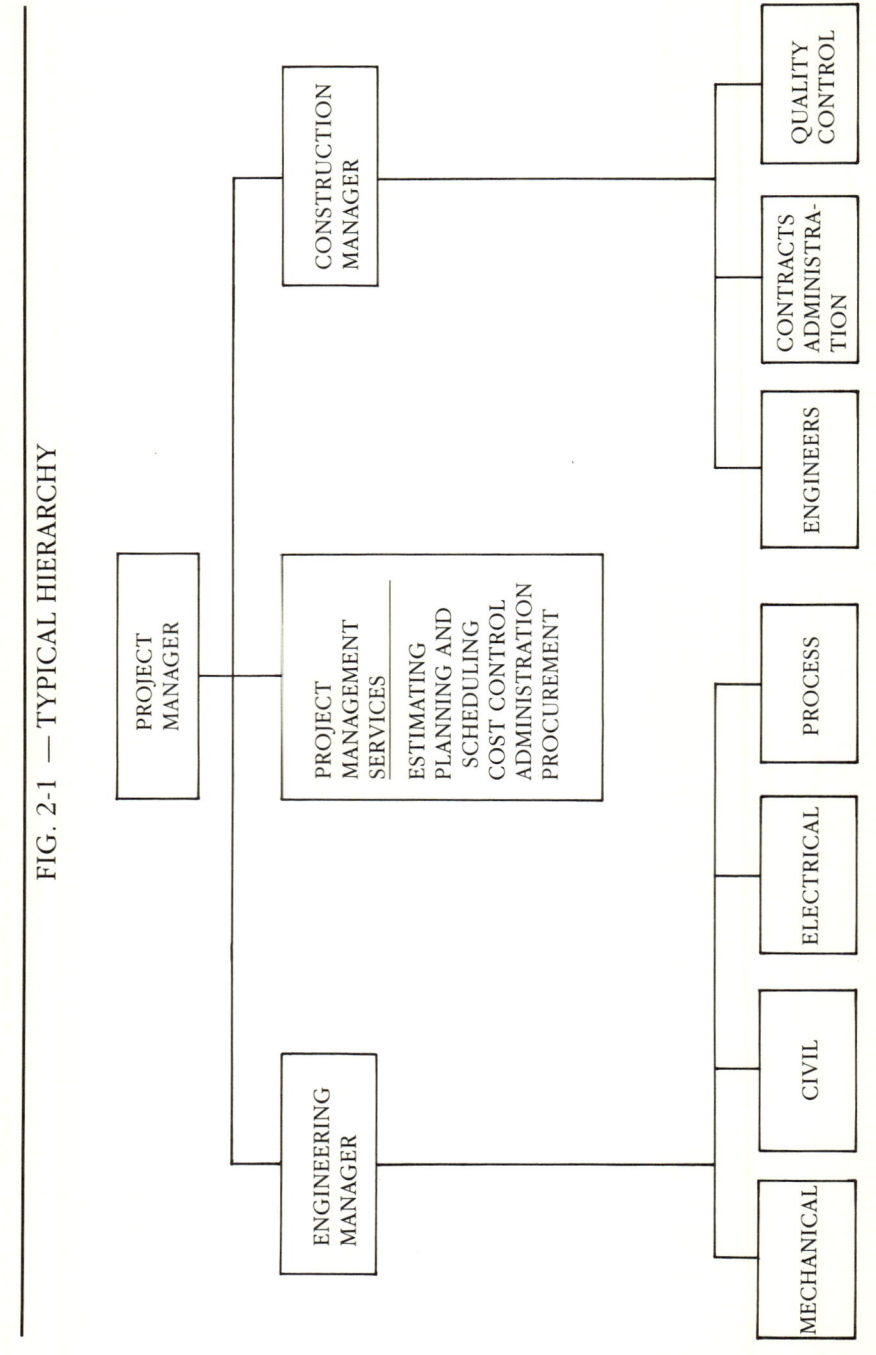

FIG. 2-1 — TYPICAL HIERARCHY

FIG. 2-2 — MATRIX ORGANIZATION

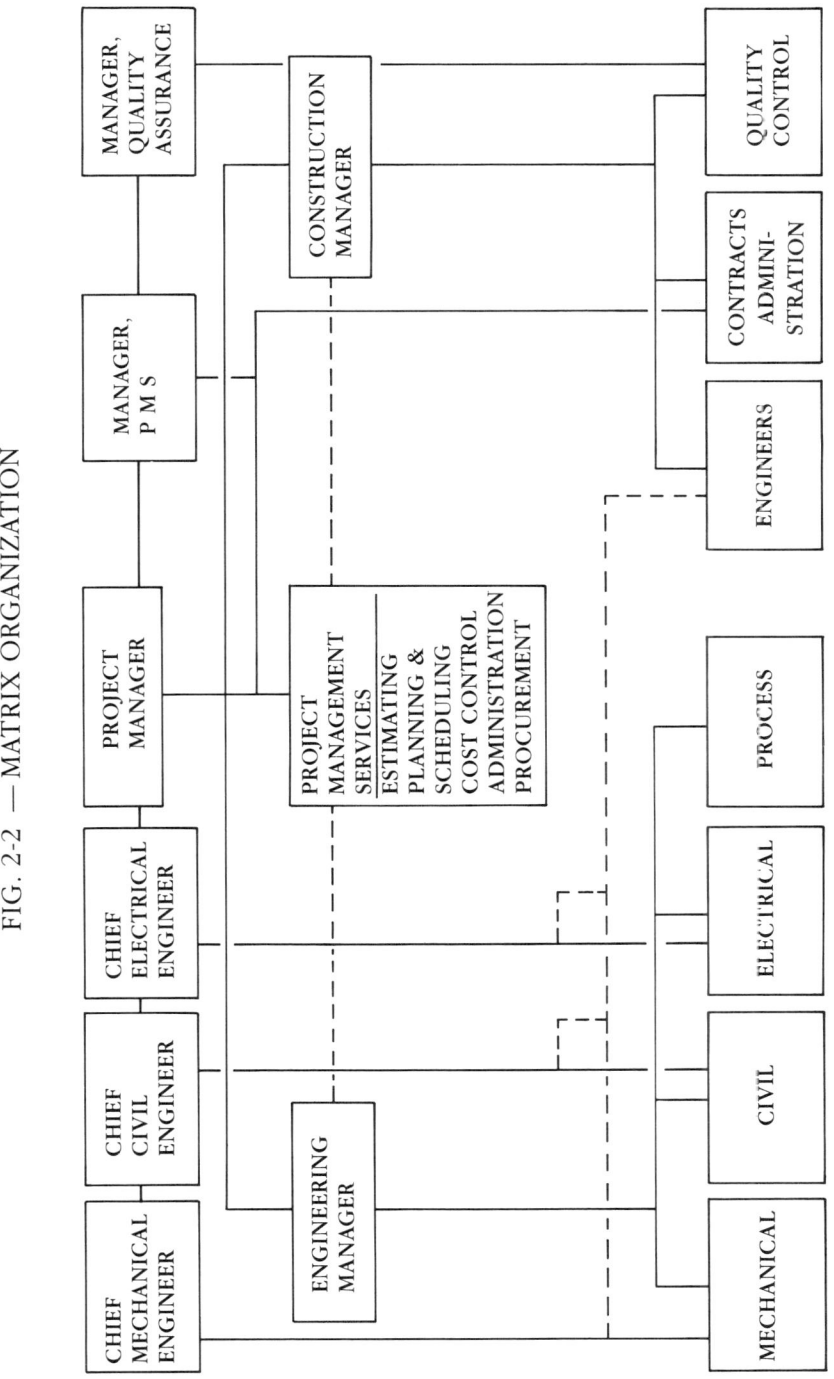

THE HIERARCHY   If a hierarchy is going to work, the principle that "the boss is the boss is the boss" must be accepted. The weakness with this idea is that the organization is only as strong as the people in it.

As an example, an engineer was once asked to trouble-shoot in a large iron ore benefication complex. There seemed to be an unusual amount of downtime. On investigation, there were some technical trends that contributed to the problem, and these were carefully catalogued so that attention could be paid to the mechanical and electrical failures that seemed to plague the company. In reality, however, the problem stemmed from the people in the organization.

A hierarchy had been set up, headed by a distinguished professorial gentleman who, recognizing his weakness as a manager, chose two strong assistant managers to run the day-to-day complex. One man was selected to cover operating personnel; the other, maintenance. When the engineer talked to operating personnel about downtime problems, the blame was put squarely on the maintenance department. In talking to maintenance people, poor operations were blamed for the downtime. What had really been happening was that the pyramid had no top. Feedback from the floor stopped at the department head, and, as a result, crosstalk between departments never occurred.

Although this example is for an operating company, the principle is the same:

In a hierarchy, a strong leader is essential.

Once that has been established, our strong leader needs to put good lieutenants and sergeants in place. If he cannot, due to reasons beyond his control, such as a simple lack of candidates, then he must be prepared to do that part of the work that is not adequately handled by his subordinate leaders.

A hierarchy leader must be capable of filling in for his subleaders' weaknesses.

THE MATRIX ORGANIZATION   The matrix organization is used in large projects. It presumes that more than one project is on the go and that the senior chiefs of each discipline have control of the technical quality of each of the projects. Sometimes these men are part of a large organization and sometimes they are consultants brought in from other organizations. A matrix can give a board of directors some assurance that things will work out technically for the better, even if the project overruns on cost and schedule.

Unfortunately, this type of organization is costly and more often than not does not achieve the technical excellence that it is thought to foster.

Why? Obviously, a matrix requires more personnel. That, in itself, is expensive, but there is a less tangible cost effect that has to do with attitudes. For one thing, if a discipline chief is responsible only for technical quality, he will hold out for a Cadillac every time, even though a Chevy may do. Why should he care? It is the project manager who is responsible for cost. And if the chief of technical quality is a stronger man, he will get his way. A subleader, on the other hand, is caught in the middle because he has to satisfy two bosses, not one, and if there are any signs of dispute, he will twiddle his thumbs while the fight goes on. What motivation has he got to proceed if the possibility exists that a change will take place?

As projects get larger or have more participants, the matrices seem to get more and more complex and farther and farther from reality. One megaproject in northern Canada, for example became so large, cumbersome, and complex that it overran its costs by double and its schedule by 50 percent. Technically, it was less than desirable, despite having had the financial resources to hire the best personnel from around the world. Why? Well, all those bored experts waiting for decisions created exotic solutions to simple problems. The project became more technically complex than it needed to be and therefore had more technical problems than necessary.

That is not to say that a matrix organization cannot be made to work. In fact, if a project manager has been selected who has no expertise in a related industry, then it is essential that a matrix be set up to ensure the technical quality of the project. Problems may arise if he does have that expertise, however, and if he has strong-minded discipline chiefs who are trying to go one way and he the other. If that is the case, then the responsibility for the budget and schedule might shift from the project manager to the discipline chiefs and the manager's role might be downgraded to that of coordinator. The odds of meeting budgets and schedules then probably would diminish.

If a strong project manager with related industrial experience has been selected, then it is essential that he control the technical input of the discipline chiefs as regards budget and schedule. This can be very difficult to do, but by involving the discipline chiefs in the budget and by scheduling processes at the beginning of the project and obtaining commitments from them at that time, the chances for success can be much greater.

WHICH TO CHOOSE?   Use a hierarchy on any size project when:

- – You are sure of your people.
- – Your project manager has experience in a related industry.
- – Tight budget and schedule control are of paramount importance.
- – Technical errors will not create disasters.

Use a matrix organization on large projects only when:

- – You are unsure of your people.
- – Your project manager has no relevant technical experience.
- – Overruns of budget and schedule can be tolerated.
- – Technical errors can result in catastrophe.

THE LEARNING CURVE   In any organization, the relative success of the project depends to some extent on a thing called "The Learning Curve," which loosely translates into "how long does it take a team to work as a team?" The more complex the team is the longer it takes to begin working as an efficient unit. By the same token, organizations tend to become more complex with time and therefore less efficient. That, in itself, is a kind of learning process, except that sooner or later, organizations seem to wake up suddenly to the fact that they are no longer efficient at all. It becomes absolutely essential, therefore, that your organization is designed to be as simple as practicable, because it will tend to become complex of its own accord.

As our chemical engineer friend found out early in the chapter, things do not always work out as planned. It is important, therefore, to monitor how the learning process is developing early in the project so that adjustments can be made. One must be wary of the tendency to solve problems by matrixing a hierarchy or overmatrixing an existing matrix. It is usually better to replace a weak link rather than support it by adding personnel, because such support only contributes to complexity. Often a better solution is to stretch surrounding personnel to fill a void. Our chemical project manager did that by taking a more active role in the design and construction management. Sometimes lateral or promotional moves can be made to cover a weakness without changing the basic organizational structure. For example, your planner and scheduler could take a more active role in cost control if the cost control chief is weak or vice versa. A strong member of a discipline team can take a more senior position, say as assistant, should the discipline head be lax in his duties.

DESIGN AUDITS AND PANEL REVIEWS   In the case where budget and schedule are important considerations, yet technical quality is paramount, a suitable alternative to a matrix organization is the staging of periodic design audits and panel reviews.

Design audits are best done by independent consultants who have no financial stake in the project. Their work need not be disruptive to the progress of the project. Primarily, their function is to prevent errors and omissions, but they can also be called upon to offer an opinion on the relative cost of the project — whether it is gold-plated or severely underestimated.

Panel reviews are usually performed as an in-house function. A panel is formed to review all aspects of the project. Its members are normally company staff who are long in the tooth, but some of the participants could be consultants.

In either case, such procedures are not welcomed by people working on a project. They feel that they are potential victims of a witch-hunt. It is extremely important, therefore, that audits and reviews be a preordained fact of life at the beginning of the project and that competent and respected individuals are selected to do the job.

Using the Right Systems

Systems, like organizations, seem to get more complex with time and have a tendency to become constipated. What systems are supposed to do is organize the flow of information so that communications are clear and events can take place in an orderly and expeditious manner.

The trouble with systems is that gray areas cannot be adequately covered. If a system is designed to sort big potatoes from small potatoes, it usually works fine for big potatoes and small potatoes but has a helluva time with medium-size potatoes. If you were to pass all of the potatoes over a screen, some of the large ones would get stuck in the holes in the screen, but many irregulars would pass through. Systems analysts drive themselves slightly bonkers trying to solve that problem instead of simply accepting the fact that systems are not and cannot be perfect.

One of the largest consulting companies in the world spent a great deal of money a few years ago designing a "perfect" system for controlling projects. The idea was based on an integrated coding system for all phases of project management — engineering, construction, procurement, scheduling, estimating, and cost control. Everything was tied to a computerized code of accounts. Theoretically, it covered practically every eventuality. But it did not work. Why?

What the system forgot to consider was that people have to work with a system, people who regard systems as secondary, at best, to the job at hand. If a complex system is introduced, it becomes a nuisance, something that interferes with getting on with the job. In the case of the company in question, an inordinate amount of time had to be spent correcting input that had originated with employees on the floor. They made errors due to the system's complexity or threw anything at the computer just to get on with the job. Since reinputting took place in the next printout, it could always be assumed that the current printout was incorrect. Once this routine got established, the printouts were rarely trusted. Though the system became more-or-less invalid, it continues to this day — garbage in, garbage out.

One must recognize that systems are inherently imperfect — always — and therefore, one must strive to *optimize* systems, not perfect them. It is better to use a simple system and deal with exceptions to it than to use a complex system to cover all eventualities.

SYSTEMS AND COMPUTERS   One highly placed executive remarked that it will not be long before a construction superintendent will be able to stroll through the site and, on spotting a potential delay in, say, pouring a retaining wall, simply walk back to his office, sit down at the computer, and punch in the deficiency. Voilà! A new, computerized CPM schedule will be printed out, complete with new costs, and the project will be up to date. Well, we can take that one step further to the voice-actuated computer. Then all the superintendent will have to do is swagger back to the office, open the door, and holler, "Get that damned retaining wall back on schedule — *on the double!*" And it will be done.

A large software company uses a US government aerospace program in its glossy brochure to promote its project management systems package. It shows sample printouts illustrating schedules, bar charts, and graphs in three colors depicting the project progress and generating forecasts. Nowhere does it say that the project depicted was overrun by 50 percent and was two years late. In short, *computerized systems will not solve your problems. They can only help to illustrate what they are.*

Project management is under great pressure to mechanize procedures, and to a large extent mechanization has been successful, particularly in the areas of accounting and cost control. Other areas have had mixed success but are improving as time goes on. A project manager must look at the project as a unit, however, and often he is faced with pressure to mechanize where it has not been done before in his organization. He runs the risk of blowing his schedule on a learning curve. As is often the case, *it is not change that is to be feared but the rate of change.*

If a hitherto uncomputerized format has been used, then it is recommended that a complete set-up *not* be introduced. Instead, begin with estimating, accounting, and cost control. Once those have been introduced to your project management program and *have been used successfully*, then and only then proceed with one more program.

If your organization has a computerized program in place, check its track record and make sure that deficiencies can be eliminated. If they cannot, then avoid the system.

BUREAUCRACIES   There used to be a saying that the amount of paper used in building an industrial plant is equal in weight to the steel used in its construction. An exaggeration, of course, but there is no doubt that the amount of paperwork generated in plant design and construction has increased, despite the arrival of the computer. Part of this may be due to the popularity of lawsuits, but it is also very probable that, like other organizations, bureaucracies increase in complexity with time. A point is reached where housecleaning must be done, and that is a very difficult thing to do.

What paperwork is designed to do is pass on information. But more often than not, it is used to pin people — or their organizations — down. "Get it in writing" is a catch phrase that means, "If you are at all serious about it, you will make a written covenant."

Certainly, when money changes hands it makes good sense to back up negotiations with written agreements. Within an organization whose people are supposedly dedicated to pulling in the same direction, however, there is something wrong with the philosophy of getting everything in writing. Nevertheless, people do have a habit of making commitments and then not following up on them. The question is: Do we make it a routine to get things in writing or do we make it an exception?

CHAPTER
3

STAGES OF A PROJECT

Projects evolve through distinct stages, beginning with an idea and following through to start up and operation. Recognizing the stages is essential to maintaining control of costs, schedule, and quality, but it is just as important to know the limitations of where you are. Project stages can be summarized as follows:

- Idea stage
- Conceptual stage
- Prefeasibility stage
- Market studies
- Feasibility studies
- Financial analysis
- Preliminary design
- Final design and construction
- Commissioning and start up
- Closing reports

IDEA STAGE

Before a project can happen, a light must go on somewhere that will illuminate the need for a project to be started or conceptualized. For example, a shortage of electrical power may lead someone to realize that a local stream could be dammed and a hydro station built to produce electricity.

Business decisions made at an idea stage are usually philosophic to the extent that the owner has to decide if this is the type of project that he would like to pursue. If it fits his overall corporate plans or goals and is compatible with the business already in hand, then he may decide to go to the next step, conceptualization.

Discussions of cost and schedule at the idea stage are normally limited to broad definitions such as, "It will probably cost in excess of a half billion dollars and will take more than three years to design and build." Obviously, if you are in the business of building 100 KVA diesel generators, you may hesitate to consider a 300 megawatt power station. And so, the idea stage establishes enough cost data to allow the management to decide whether or not to follow up on the idea.

In corporations, ideas are not usually presented to the board in a formal way but are sometimes mentioned to get a reaction, particularly if some money is likely to be spent in the next stage. Ideas can begin with anyone in the organization. A successful enterprise will encourage input from every level. It never hurts to listen.

CONCEPTUAL STAGE

Once an idea has shown merit, it becomes necessary to establish the shape of the project, to get a better feel for its scope and its size. In our hydro project example, some expertise needs to be introduced. Can a dam be built on the site? Are the terrain and soil conditions compatible? How high would the dam be? How much land would be flooded? How much power could be produced? What infrastructure is required?

It is not uncommon at the conceptual stage to develop concepts that are grandiose or inadvertently frugal, especially if a high degree of expertise has not been introduced or if some of the key facts are missing. As a result, some reconceptualizing should be expected during the prefeasibility stage.

Costs can be put to a concept. However, the accuracy of an estimate may be wildly out, probably in the range of ± 50 percent. At this stage, an estimate's usefulness is in establishing an idea of the cost commensurate with the aims of the project and the company. It is usual in establishing capital and operating costs to factor them from similar projects. For example, both capital and operating costs could be estimated for our hydro project on a cost per unit of power output, cost per foot of dam height, cost per volume of water, or a composite or weighted average of all of these.

When a concept has been established there is normally a flush of enthusiasm associated with it. Beautiful drawings are made, photographs are taken, and sometimes models are made. There is a dog-and-pony show given, a film is shown, and a band plays up-beat music as the credits come on the screen. Then some guy in the back row asks, "Is it financially feasible?" And reality begins to set in.

PREFEASIBILITY STAGE

On large projects, feasibility studies are expensive, and so a prefeasibility study is sometimes carried out in order to obtain funds to proceed to the next step. In the case of mining projects where an ore body is not yet proven, a prefeasibility study is sometimes done in order to raise funds for drilling to prove up enough ore to warrant carrying out a feasibility study on mining and milling.

One of the main functions of a prefeasibility study is to solidify all of the pertinent data into a single document. It is surprising how many misconceptions developed during the conceptual stage are brought to light by a prefeasibility study. In one sense, it is the formalization of the conceptual stage.

What a project manager should attempt to do when writing a prefeasibility study is to prepare a report that has all of the essential ingredients of a feasibility study, except that the accuracy of the cost data is limited by the fact that no detailed engineering or estimating has been done. It is tempting to leave out data where there is none, but the true professional makes a best guess and qualifies it as such, leaving out nothing that cannot be estimated, however wild that estimate may be. In essence, he tries to document all of the facts, even though he may not have them. He must recognize that even a poor guess is valuable if it raises a question later on at the feasibility stage and leads the project team to the right answer in the end. Therefore, if you are writing such a report, remember that the easiest way to get the right answer is to throw the wrong one on the table and defend it to all who challenge you. That takes guts, but it gets results.

A prefeasibility study covers the following main points:

- Introduction
  a summary of what and who prompted the report, its purpose and scope
- Location of the project, with maps and references
- History of the project
  previous studies, if any
  what occurred in the idea stage
  what occurred in the conceptual stage (or stages)
- Process
  its history
  its validity, including precedents and tests
  its application and description
- Design
  description
  drawings
- Infrastructure, as applicable
- Schedule
  bar chart is sufficient, including proposed feasibility studies, decision times, design, and construction
- Capital cost estimate, including the basis of the estimate
- Operating cost estimate
- Market analysis
- Financial analysis
- Conclusions
- Recommendations
- Bibliography

There is always a tendency to short-cut the prefeasibility study. But you should assume that a ''hot'' market could prompt a rush into design and construction, and, despite the risks, the job could go ahead on the basis of the prefeasibility study. So the less left out the better.

In projects where a process is involved, it is usual for only theoretical work or bench-scale testing to be done at the prefeasibility stage and for part of the work anticipated, such as pilot plant testing, to occur as a result of the report.

Enough engineering design is needed to properly estimate the capital and operating costs. Essentially, building sizes and construction techniques are described. It is sometimes important to obtain soils tests for foundations, especially if unstable or difficult conditions are expected.

Estimates used in a prefeasibility study are order-of-magnitude (see Chapter 10), which have accuracies in the range of +25 percent to 0 to as high as +40 percent to 0. Costs are a combination of telephone quotations from equipment suppliers and contractors and costs factored from similar projects. Schedules are usually drawn from experience and modified by long-delivery items or extraordinary circumstances such as remote locations.

Some attempt should be made to analyze the market. However, it should be recognized that a full-scale market study is normally required at the feasibility stage. And finally, a financial analysis indicating rate-of-return and payback period should be provided.

MARKET STUDIES

If there is a shortage of 30 megawatts of power, you may be foolish if you build a 30-megawatt power station. What if the demand goes up? Or down? What is going to happen to the prices? How can you know ahead of time?

Market studies may be crystal-ball gazing at best, yet they are absolutely necessary if financing of a project is going to be sought. While there is no such thing as a sure thing, credibility is the name of the game. Economists have a history of wrong guessing and so do people who do market studies. Nevertheless, if you are going to take a risk on the economy, who are you going to ask for advice? Gut feelings carry no weight with banking institutions. You had better hire a reputable market analyst, preferably as a consultant.

Elements of a market study can be listed as follows:

- History
- Statistics
- Competition
- Future competition
- Local and world economy
- Trends
- Recommendations

What cannot be foretold in a market study is what usually happens. A void in the market place stimulates a demand and sends prices up. Company A does a market study and feasibility study, presents its findings to Banker Z, who finances the project. Meanwhile, Company B goes through the same exercise, obtaining financing through Banker Y. Depending on the demand stimulation, a whole alphabet of companies and bankers can get into the act, each not knowing of the other, and before long the market is oversupplied and prices plummet. Just such a situation existed in the early 1960s, when pulp and paper shortages stimulated the simultaneous construction of over twenty mills around the world. The result? Prices dropped and so did some mills — dropped out of business, that is. While a market study cannot predict such events, it should make some attempt to deal with them. The real tool to use in meeting such eventualities, however, is the feasibility study.

FEASIBILITY STUDIES

Even though you as an owner may lead the world in expertise and have a fine track record of successful projects, even though you may have internal financing and fully intend to design and build your own plant, it is always better to hire an independent consultant to do your feasibility study. Although the benefits of using a consultant may be many or few, the best reason for hiring him is that he must view the project at arm's length. He must be objective in his approach if he is at all reputable and intends to remain that way. Too often we are victims of our own ambitions, having our thinking clouded by our desires. An independent consultant may not escape your influence entirely, but he does buy you credibility with those who control the purse strings.

Feasibility studies are produced in neatly bound books with illustrations and drawings designed to impress their readers even if the results are negative (which they rarely are, since the project gets stopped at the first indication that it is not feasible). The cost of producing the study varies with the amount of detail required and the capital cost of the project. Invariably, the thickness of the feasibility study is directly proportional to the cost of the study. After all, people reason, what do you get for your tens of thousands of dollars? A few books. So they damned well better be big! One stainless steel mill project came up with twelve volumes varying in thickness from one and a half to four inches! A closer inspection revealed that the volumes had been padded with reams of specifications — stuff that should have remained in the filing cabinets of the consulting group. Who, one may ask, is kidding whom?

What a feasibility study is meant to do is describe the project and define its scope to the extent that the costs and financial analysis make sense. Its costs and their accuracy reflect the amount of work done, but all of the work done need not be included in the report. It should only be referred to so that, if need be, it can be retrieved from the files for examination.

The format of the report should be essentially the same as for the prefeasibility study. In fact, if the prefeasibility study has been done properly, it should serve as a basis for the consultant's report. The main differences are in the amount of work done in preparation for each section and the resultant accuracy. A feasibility study is characterized by the following:

- Process work is usually the result of pilot plant work or similarly exhaustive studies by an expert in the field.
- Engineering and design are done in sufficient detail to obtain written quotations from suppliers for equipment and from contractors for buildings and installation. The likelihood of major design changes should be considered low at this point.
- Capital cost estimates are made to an accuracy of $+15$ percent to $+20$ percent, -5 percent and are in sufficient detail to serve as a budget.
- Operating costs are made to a similar accuracy and can also serve as a budget.
- If a separate market study is done, its results are reported in the feasibility study.
- Planning and scheduling are done in sufficient detail to begin and to define the life of the project.
- A financial analysis is done, showing cash flows, rates of return, payoff periods, and breakeven, with a sensitivity analysis that indicates which factors affect the financial success of the enterprise and to what extent.
- Environmental impact, if applicable, is discussed.
- Conclusions and recommendations are included.

Since feasibility studies are often read by nontechnical people who are reluctant to plow through the material, it is fashionable to write an "Executive Summary" that will serve the dual purpose of encapsulating the contents of the study and stroking the egos of its readers. As insolent as that may sound, the summary does serve a useful function, not just for the executive, but for anyone who is interested in a quick overview.

FINANCIAL ANALYSIS

Accounting and financial analysis seem to mystify most engineers and project management team members to the point where many of them take an MBA in order to get demystified. It may be helpful to you to have an overview of popular buzz words, definitions, and methods of analysis.

The simplest method of getting an idea of the viability of a project is to calculate the payback period. This is merely a calculation of how long it takes to pay back a capital investment, not allowing for the value of money. Suppose, for example, you invest $10 million to build a plant, and its gross income is $5 million per year. It costs $3 million to operate and pay its taxes, netting $2 million per year.

$$\frac{\$10 \text{ million}}{\$2 \text{ million/year}} = 5 \text{ years to pay back the } \$10\text{-million investment}$$

That $2 million per year is known as cash flow.

An overall average rate of return (ROR) or return on investment (ROI) can also be simply calculated if the life of the project is finite. But this calculation does not take into account the value of money either, nor does it take into account variations in cash flow. In our example, suppose the plant life is ten years.

$$\text{average cash flow} = \$2 \text{ million/year}$$

$$\text{less depreciation and amortization} = \frac{\$10 \text{ million}}{10 \text{ years}} = \$1 \text{ million/year}$$

$$\text{ROR} = \frac{\text{average cash flow} \times 100}{\text{investment}} = \frac{\$2 \text{ million} \times 100}{\$10 \text{ million}} = 20\%$$

$$\text{ROI} = \frac{\text{average net income} \times 100}{\text{investment}} = \frac{\$1 \text{ million} \times 100}{\$10 \text{ million}} = 10\%$$

Our example assumes equal cash flow. However, if it were to vary to, say, $3 million per year in the first two years, but decrease to $2 million for the next six years, and then continue at $1 million per year for the rest of the project, then the overall ROI and overall average ROR would remain the same, but the cash would be put into the kitty sooner. The project, therefore, would be safer to the investor, especially in an unstable economy. Such is the weakness of the ROR/ROI method.

The ROR/ROI analysis, viewed in conjunction with the payback period, is an indicator of project viability. Things get complicated but more accurate when the value of money is considered. Two popular and similar methods of calculation used are the discounted cash flow (DCF) method and the net present value (NPV) method. The NPV method is considered slightly more conservative. But in general, if the difference in the results of the two methods is sufficient to sway your project one way or the other, then you should not consider the project at all!

Discounted cash flow return on investment (DCF/ROI) calculations assume an ROI as an interest rate, and the answer is expressed as an amount. A trial-and-error calculation is done using several assumed ROIs until the answer is equal to the investment amount. A computer should be used to do this if you require an accurate percentage.

In our example, we know from ROR/ROI calculations that the percentage is in the 10 to 20 percent range. Present values, extracted from interest tables, are as follows: $PV = \dfrac{1}{(1 + i)^n}$

where i = interest and n = number of periods.

PRESENT VALUE OF $ 1

Year	10%	15%	20%
1	.909	.870	.833
2	.826	.756	.694
3	.751	.658	.482
4	.683	.572	.402
5	.621	.497	.335
6	.564	.432	.279
7	.513	.376	.233
8	.467	.327	.194
9	.424	.284	.161
10	.386	.247	.135

If we try the calculation at varying interest rates, we get the results shown on page 36:

Year	Cash flow $ MM	Discount factor 10%	Present value	Discount factor 15%	Present value	Discount factor 18%	Present value	Discount factor 18.5%	Present value
Now	−10	1.000	−10	1.000	−10	1.000	−10	1.000	−10
1	+3	.909	2.73	.870	2.61	.847	2.54	.844	2.53
2	+3	.826	2.48	.756	2.27	.718	2.15	.712	2.14
3	+2	.751	1.50	.658	1.32	.609	1.22	.601	1.20
4	+2	.683	1.37	.572	1.14	.516	1.03	.507	1.01
5	+2	.621	1.24	.497	0.99	.437	0.87	.428	0.86
6	+2	.564	1.13	.432	0.86	.371	0.74	.361	0.72
7	+2	.513	1.03	.376	0.75	.314	0.63	.305	0.61
8	+2	.467	0.93	.327	0.65	.266	0.53	.257	0.51
9	+1	.424	0.42	.284	0.28	.225	0.23	.217	0.22
10	+1	.386	0.39	.247	0.25	.191	0.19	.183	0.18
Net present value			+3.22		+1.12		+0.13		−0.02

It can be seen on page 36 that the DCF/ROI is slightly less than 18.5 percent. An accurate interest rate could have been calculated by further trial-and-error or by computer to give a net present value of zero.

The net present value (NPV) method is similar to the DCF method, except that present value of the project cash flow is calculated at a desired, specified interest rate. The difference between the present value and the investment amount is the NPV. If the answer is negative, the project is not feasible at the desired interest rate. If it is positive, it may be compared to the cost of capital to render a go/no-go decision.

The DCF method will calculate the rate of return of the project and does not consider the desired rate, while the NPV method specifies the rate and evaluates the excess or deficiency of cash at the end.

A sensitivity analysis usually is included in a financial analysis and may cover many factors, such as price of product or operating costs, that could change with time and could affect the feasibility of the project. Figures 3.1 and 3.2 are examples of a cash flow statement and sensitivity analysis for an industrial mining project.

PRELIMINARY DESIGN

One of the primary objectives of distinguishing preliminary design from final design is to establish a definite time frame for freezing the process and plant design so that cost and schedule upsets are minimized. Projects that have continual scope changes are not controllable. When basic concept changes are made in the middle of a project, then cost control becomes a joke. It is essential, therefore, that a line be drawn which indicates that preliminary design is complete and major scope changes are not permitted.

A second objective of freezing the design is to allow preparation of a definitive estimate to an accuracy of 10 to 15 percent to control the job. In order to do that, about 20 percent of the total engineering must be done, as measured by completion of drawings. Most often this also requires that commitment be made to equipment manufacturers to the extent that their equipment sizes and designs affect the layout of the plant facilities and building sizes.

Unless a firm resolve to harness the team is made throughout

FIG. 3-1 — CASH FLOW STATEMENT = BASE CASE

	1983	1984	1985	1986	1987	1988	1989
Tonnes of ore	–	75,000	150,000	150,000	150,000	150,000	150,000
Recovered grade (%)	–	6.8	6.8	6.8	6.8	6.8	6.8
Recovered product (tonnes)	–	5,100	10,200	10,200	10,200	10,200	10,200
Price of product ($)	–	900	900	900	900	900	900
Gross revenue	–	4,590	9,180	9,180	9,180	9,180	9,180
Total operating cost	–	2,940	5,430	5,430	5,260	5,260	5,260
Operating profit	–	1,650	3,750	3,750	3,920	3,920	3,920
Quebec mining tax	–	–	–	–	115	231	270
Quebec income tax	–	–	–	–	74	101	89
Canadian federal taxes	–	–	–	–	274	380	532
After tax profits	–	1,650	3,750	3,750	3,457	3,208	3,029
Total capital expenditure	2,200	5,960	–	2,900	–	–	1,000
Net cash flow	(2,200)	(4,310)	3,750	850	3,457	3,208	2,029
Present value of NCF @ 15%	(2,200)	(5,950)	(3,111)	(2,549)	(570)	1,028	1,907
Present value of NCF @ 30%	(2,200)	(5,518)	(3,296)	(2,907)	(1,695)	(829)	(408)
Internal rate of return	–	–	–	–	9.9	22.1	26.5
Equity cash flow	(2,200)	(4,310)	3,750	850	3,457	3,208	2,029
Equity present value @ 15%	(2,200)	(5,950)	(3,111)	(2,549)	(570)	1,028	1,907
Equity rate of return	–	–	–	–	9.9	22.1	26.5
CAPITAL EXPENDITURE DETAILS							
Major milling assets (28)	2,200	4,480	–	–	–	–	–
Preproduction	–	550	–	–	–	–	–
Development	–	930	–	1,450	–	–	1,000
Ongoing mining assets (10)	–	–	–	1,450	–	–	–
Total capital expenditure	2,200	5,960	–	2,900	–	–	1,000

NOTE: All figures are $ × 10^3, except where indicated otherwise.
(Courtesy of Golder Associates)

the engineering group, changes quite often will continue ad infinitum. While one must be leary of steamrolling a design freeze prematurely, one should be just as cautious about preventing people from continuously changing their minds.

FINAL DESIGN AND CONSTRUCTION

It is not uncommon for construction to begin even before preliminary design has been completed. Ground has been broken and concrete poured in some rare cases before it is known precisely what is going to be installed. It is strictly a question of the pressure of competition or the strength of the urge to produce revenue. Normally though, construction begins at the latter stages of preliminary design at the earliest.

Engineering design and construction are inseparable, even though two distinct groups normally perform the work. While it

1990	1991	1992	1993	1994	1995	1996	1997	1998	1999	TOTAL
150,000	150,000	150,000	150,000	150,000	150,000	150,000	150,000	150,000	150,000	2,325,000
6.0	6.0	6.0	6.0	6.0	6.0	6.0	6.0	6.0	6.0	–
9,000	9,000	9,000	9,000	9,000	9,000	9,000	9,000	9,000	9,000	–
900	900	900	900	900	900	900	900	900	900	–
8,100	8,100	8,100	8,100	8,100	8,100	8,100	8,100	8,100	8,100	131,490
5,170	5,170	5,170	5,170	5,170	5,170	5,170	5,170	5,170	5,170	81,280
2,930	2,930	2,930	2,930	2,930	2,930	2,930	2,930	2,930	2,930	50,210
321	321	321	321	321	321	321	321	321	321	3,831
94	117	118	119	119	120	120	120	120	120	1,433
491	752	790	790	790	790	790	790	790	790	8,751
2,024	1,740	1,701	1,700	1,700	1,699	1,699	1,699	1,699	1,699	36,204
–	–	–	–	–	–	–	–	–	–	12,060
2,024	1,740	1,701	1,700	1,700	1,699	1,699	1,699	1,699	1,699	24,144
2,667	3,234	3,717	4,136	4,500	4,817	5,093	5,332	5,541	5,722	5,722
(86)	127	287	410	505	577	633	677	710	735	–
29.3	30.9	32.0	32.8	33.3	33.7	34.0	34.1	34.3	34.4	–
2,024	1,740	1,701	1,700	1,700	1,699	1,699	1,699	1,699	1,699	24,144
2,667	3,234	3,717	4,136	4,500	4,817	5,093	5,332	5,541	5,722	–
29.3	30.9	32.0	32.8	33.3	33.7	34.0	34.1	34.3	34.4	–
–	–	–	–	–	–	–	–	–	–	6,680
–	–	–	–	–	–	–	–	–	–	550
–	–	–	–	–	–	–	–	–	–	3,380
–	–	–	–	–	–	–	–	–	–	1,450
–	–	–	–	–	–	–	–	–	–	12,060

is theoretically possible to complete all of the engineering before beginning construction, such cases are rare. One large aluminum producer tried it in the early 1970s when lack of funds delayed the start of construction. Instead of making hard purchases of equipment, they issued letters of intent in order to get the engineering information. The project suffered from lack of motivation among personnel, however, and the budget and schedule went out of sight. And even though the design had been considered complete, site conditions required design modifications. The engineering never really ceased.

COMMISSIONING AND START UP

A commissioning stage can be defined as a check-out period prior to starting up the plant. Equipment has been checked out by the suppliers and/or engineers and signed off as ready for operation.

FIG. 3-2 — SENSITIVITY ANALYSIS

CASE	Rate of Return (%)	NPV at 15% ($10³)	NPV at 30% ($10³)
BASE	34.4	5,722	735
Product $1,100	59.1	13,000	4,437
Product $1,000	46.8	8,968	2,676
Product $850	28.4	4,051	(283)
Product $800	22.1	2,217	(1,396)
Product $750	15.8	263	(2,581)
Product $700	9.3	(1,886)	(3,845)
Capital costs increased 10%	30.5	5,138	98
Capital costs increased 20%	27.2	4,488	(582)
Capital costs increased 30%	24.9	3,916	(1,130)
Operating costs increased 10%	29.6	4,579	(66)
Operating costs increased 20%	22.5	2,311	(1,322)
Capital +20%, operating +10%	22.2	2,697	(1,660)
Product $800, capital +20%, operating +10%	11.8	(1,261)	(4,060)
Product $750, capital +20%, operating +10%	6.0	(3,512)	(5,378)

(Courtesy of Golder Associates)

Pumps turn, conveyors are proven to run forward instead of backward, alarms work, interlocking wiring has been checked, and compressors produce compressed air. Sometimes the entire plant is run for a period of time without material, and sometimes partial load conditions are tested or simulated.

Depending on the product handled, start up usually begins on a reduced output basis, gradually increasing until plant output capacity is reached. This stage is sometimes handled by a start-up crew who are formed from a combination of the commissioning team, the engineers, and the operators.

CLOSING REPORTS

Human nature makes it difficult to finish construction, complete commissioning, or walk away from start up. There always seems to be something else to do. When you buy a car, the warranty very definitely spells out the end of the manufacturer's responsibility. Similarly, in project management it is necessary to clearly define the end of the work on an official basis. This is done with closing reports.

CHAPTER

4

WHEN TO USE
A CONSULTANT

CASE 1 — FOR TECHNICAL EXPERTISE

A large consortium began working on a megaproject more than two decades ago. As the work they were doing paralleled that of others, and since it was considered to be in an area of new technology, there was a great deal of stress put on security. A research laboratory was set up, and the finest talent available was hired to do original process work. Although the basis for the process was in the public domain, some kind of breakthrough was needed to give the project the stimulus it needed to warrant financing of such a large enterprise.

No expense had been spared on personnel and equipment. Bench-scale testing led to pilot plant work. Although results were not absolutely conclusive, a similar commercial plant owned by competitors had begun production, lending a great deal of credibility to the project. The need for secrecy, however, became more acute in the eyes of the consortium, so it was decided that all engineering, up to the end of the feasibility study, would have to be done in-house. Personnel hired for the engineering could be used to direct the project beyond the feasibility stage and would probably go on to run the operations, thus justifying a fairly large staff. Thirteen years and $33,000,000 later, the consortium felt that they were ready to call in a consulting company to do engineering, procurement, and project management. The feasibility study was complete and the conceptual design of the plant was considered a *fait accompli*.

When a consulting company was engaged they were not privy to the details of the work that had gone on, but they were experienced in the field. What they found after signing appropriate secrecy agreements made their collective eyes roll.

On the process side, a surprising number of assumptions had been based on interpolated data. Many of the curves had been drawn by computer from points plotted from pilot plant readings. Some results looked as though they had been shot onto the paper with a scatter gun. It was obvious to the consultant that more process work needed to be done.

Feasibility had been based on a capital expenditure forecast of $500,000,000. The plant design used, however, was hopelessly inadequate, containing oversights obvious to the practical eye. When the consulting company tried to gently suggest rectifying some of these problems, the client became quite defensive. After all, he had spent $33,000,000 and thirteen years of effort on the project and had, in fact, progressed to being one of the leading authorities in the business. The consultant, on the other hand, had only been on the project for a couple of months. What did he know?

At least, said the consultant, let us check the capital cost. Only, said the consortium, if you do it on our design. So, capitulating, the consultant agreed. The capital cost estimate came out at $750,000,000! It took quite some time to get the consortium to agree that their own in-house estimate had been out by 50 percent and even more time to convince them to do more pilot plant work and a complete new design. With a little help from inflation, the project finally came in at $1,700,000,000!

In an overview of the entire program and with the advantage of twenty-twenty hindsight, one can easily conclude that a lot of grief may have been saved had a consulting team been brought in earlier. It is absolutely clear that the consortium, contrary to their own belief, did not have personnel who were experienced enough to control the project. Despite having spent a lot of money, they brought in the wrong people.

CASE 2 — FOR MANPOWER PLANNING AND LEVELLING

Some companies do have experienced project people on staff, yet they still manage to run into cost and scheduling problems. Usually these problems come about when internal expansion and acquisition activities reach high levels. Project people get stretched too thin and the inevitable results are overruns in costs and schedule. Overall management of the capital spending program can take care of that.

In order to control a capital spending program it is necessary to do a periodic summary of anticipated capital costs and durations and from those estimate your manpower requirements. Costs of project management usually run in the range of 8 to 20 percent of the capital costs, including engineering, procurement, and construction management and depending on the size and complexity of the project. For estimating purposes, an average of 15 percent can be used. If you want to estimate the average manpower requirements for a $50,000,000 plant taking two years to build, then

$$\text{project management cost} = 15\% \text{ of } \$50 \text{ million}$$
$$= \$7,500,000.$$

If average weighted salaries are in the range of $30,000 per annum in our case, then we must add about 80 percent for burdens and overheads. The average salary then becomes

$$\$30,000 + .8\,(30,000) = \$54,000.$$

Manpower requirements are, therefore, as follows:

$$\frac{\$7,500,000}{54,000} = 139 \text{ people over two years (or 69 employees/year).}$$

It is unrealistic to put sixty-nine people on the job right at the beginning and to take them all off right at the end of two years.

So an estimate of manpower levelling must be made. Actual projects usually staff and shed people on a form of bell curve (Figure 4.1), and for the purpose of estimation, the peak will exceed the average by about 50 percent. Our peak manpower requirements, therefore, will be about 1.5 × 69 = 104 persons.

Naturally, when the job actually begins, the number of people will not be added to fit a precise bell curve. The curve will take on a modified form (Figure 4.2), which will change in accordance with the quality and experience of personnel and the accuracy of the estimate.

While the above example is a simplification of the problem, it could be expanded to cover several projects that are proposed or in progress. The aim of this method is to help the owner decide if he has enough people on staff to tackle the projects at hand. Often this simple exercise will lead him to hire more people or to

FIG. 4-1 — MANPOWER DISTRIBUTION CURVE

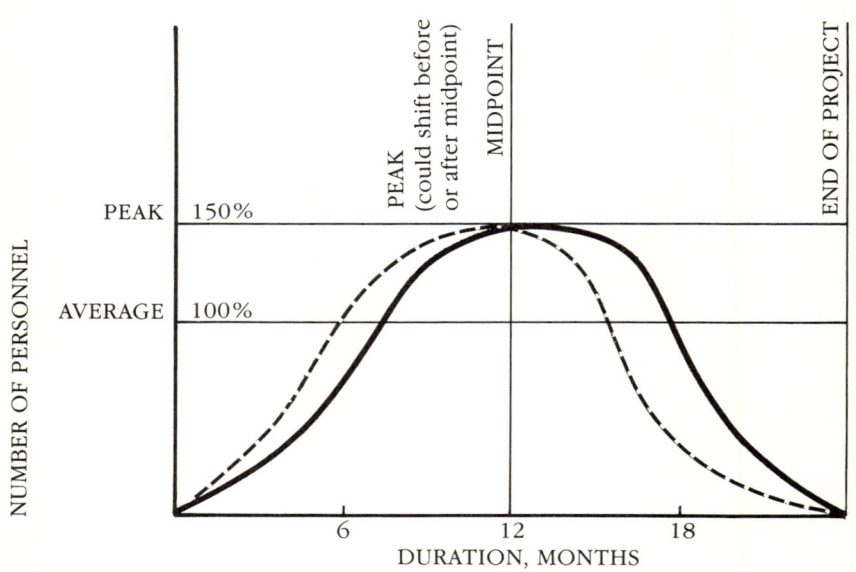

engage a consulting firm to take the workload off his shoulders. The decision to take a consultant frequently is based on the fear of building a large staff and then having to shed people as the project or projects taper off.

CASE 3 — FOR CREDIBILITY

If you intend to go for financing, either internally or with an outside financial institution, you may find some reluctance if you have prepared your own feasibility study, especially if the predicted rate of return is marginal. People who put up money will do so only after they have run out of reasons to say no. Lack of credibility may be one of those reasons. Having a consultant do your feasibility study will give you this credibility, provided that you choose one who is reputable.

FIG. 4-2 — MODIFIED MANPOWER DISTRIBUTION CURVE

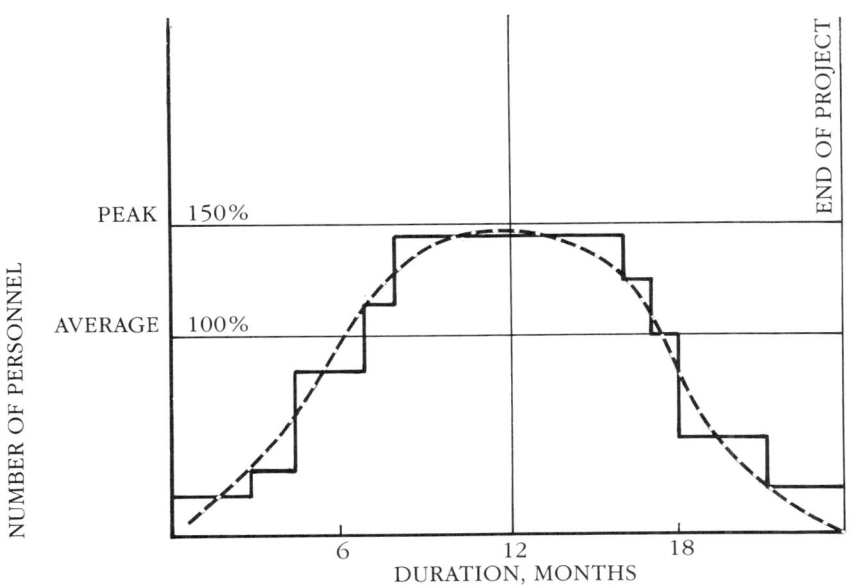

CHAPTER
5

HOW TO CHOOSE A CONSULTANT

HOW A CONSULTANT MAKES HIS MONEY

In Chapter 4 we talked about when you should use a consulting company. In summary, we said that you should go to a consultant for his expertise and when you simply do not have the manpower to do the job yourself (or if you can get that kind of manpower, but don't know what to do with it when your job is done). Also, engaging a reputable consultant to do your feasibility study will give you credibility.

People who are unaccustomed to working with consultants are usually shocked when they first see their rates. In all probability you will feel that they are two to five times what common sense says the rates should be. The shock takes some getting used to, and you may wonder, "What makes a man worth that much anyhow?"

Very specialized and unique individual consultants or companies with patented processes and inventions can and should command high rates. All rates need to be competitive, of course, commensurate with the degree of skill offered and the current economy. All levels of consultants are selling the same basic commodity, however — the manhour. While there are a variety of methods of calculating the fee for manhours, certain basic costs need to be covered over and above salaries. These are:

- Fringe benefits and payroll costs
- Overhead costs
- Management overhead costs
- Availability allowances
- Profit

Fringe benefits and all that goes with them are sometimes known as a salary burden. Such items as payroll costs, insurance, vacation pay, dental plan, and medical plans fall into this category also. Costs vary from 20 to 30 percent of the basic salary.

Overhead costs include building space, utilities, advertising, telephone, and telex. Costs vary from 30 to 50 percent of salaries.

Management overhead includes the cost of general management of the consulting firm and generally runs around 10 percent of basic salary costs.

When a consulting company hires individuals, it cannot assume that their services will be paid for by clients on a full-time basis. There has to be a cost built-in for availability. From good to bad times, this can amount to 20 percent or more.

And finally, profit, or more precisely, the earnings mark-up or fee that depends so much on the value to the user can vary from a negative percentage to several hundred percent of basic salary costs.

Taking the above information, breakeven for a typical consulting firm is in the following range:

Burden	20 to 30 percent
Overhead	30 to 50 percent
Management overhead	10 percent
Availability allowance	20 percent
Basic breakeven =	80 to 110 percent over salaries

Typically, profit is in the range of 15 percent, but you may wonder about multipliers of 2.5 to 3.0 or even more times salaries. The higher the mark-up, the more able the consultant to retain quality personnel on stand-by, bearing in mind that the right people is the number one criterion for a successful project. Keeping them together as a team is also important, otherwise you will pay for the learning curve by the total number of manhours used. Higher mark-ups also allow the consultant to delve into research and to expend more effort in business development, thus increasing his capabilities and reliability. In short, you get what you pay for.

Rogues Row — What Bad Consultants Can Do to You

During the bidding/negotiating stage, any consulting firm will put its best foot forward. You will be presented with glossy brochures showing very impressive projects and happy, energetic individuals wearing hard hats bearing the company logo. They will have blueprints in hand and will appear inordinately intelligent. There will be fractionating towers in the background. Office scenes will show similarly intimidating individuals in conference around a drawing table or examining a CPM network on the wall or watching someone who is obviously a tad more academic standing before a blackboard with $E = mc^2$ scrawled on it. Always they will be in shirtsleeves and ties, hard at it. In the lab, white-coated individuals are shown bent over microscopes or staring expectantly at a distillation column. One company couldn't resist the temptation to show a gorgeous geophysicist taking rock samples in Arizona, her golden hair flowing from beneath a stetson, her designer jeans a curvaceous complement to the rugged striations of the hills. This is not to say that a well-prepared brochure should be condemned. Indeed, if a brochure is needed, then it should be as appealing as possible and should be considered as one measure of the consultants' abilities.

Sometimes projects completed are described in the brochures; at other times they are listed on a separate document. One company claimed expertise in metallurgical complexes and displayed photographs of completed projects and of work in progress. In a separate section, they proudly displayed a chemical plant for which they had "total EPC responsibility." On checking out these claims, it was discovered that they had, indeed, worked on the metallurgical plants. They had been a subconsultant, however, to another company that really did have metallurgical plant expertise. The service the consultant provided was civil engineering, a discipline that need not have particular knowledge of metallurgical plants but only of heavy industry. Yet there it was in print — they were genuine experts in building metallurgical plants! As for the chemical plant, the consultant did the EPC work all right. But the client had to step in to manage it due to the consultants' incompetence, finally kicking them off the job when it was 90 percent completed.

Consultants who are shy on ethics are fond of presenting a slate of talented candidates who may or may not be assigned to your project. You may very well be under the impression that they will be yours exclusively. Chances are they will be chasing other work or spending a token amount of time on your project, depending on the company workload.

Speaking of workload, when such a consultant is short on work, you can bet that your project will become a dumping ground for every derelict in the office, including the most senior — and expensive — staff around. In the opposite case, when work is a-plenty, you could find the company dregs on your roster, and even those will be spread thinly. Either way, you can't win unless you change consultants.

One consulting group found themselves oversold during an inflationary period when personnel were hard to find. What they could bring in off the streets either left a lot to be desired or came at an exorbitant premium. Everyone was expected to carry extra workloads, but of course the economy being inflamed meant that productivity had ebbed. A project manager got stretched to handling three small projects at the same time, each of which was undermanned. What was worse, each of the clients liked to drop in to inspect progress on a more-or-less regular basis.

Now, in order to maintain some semblance of productivity and coordination, our intrepid manager took to working all personnel on a single project a week at a time, leaving the other two to idle. An untimely visit by the wrong client meant removing one set of drawings from the drafting boards and replacing them with another so as to make it appear that the client's project was in progress. One would expect that sooner or later two or more clients would show up at the same time. It never happened. But the team did put up the wrong drawings one time, spilling the beans. It was too late to change consultants. The client simply had to grin and bear it — and increase the frequency of his visits.

The name of the game is selling manhours. If that is a consultant's prime objective, he will come at you with all sorts of experts armed with the spectre of what will happen to your project if you do not add them to your job. This will occur after the project contract is signed, of course, and often after a working budget is struck. Naturally, that means a change order. And if the consultant's company is heavily matrixed, then the experts offered have to be monitored by *senior* experts, who will contribute manhours to the job till the cows come home. Before you know it, a virtual army of staff will descend upon your project. Never fear, however, because your budget will not be blown — it will be changed!

One enterprising consulting firm contracted a rather large order for shop-fabricated low-pressure vessels to a supplier, who in turn subcontracted to a fabricator. The supplier selected the fabricator on the basis of price and competence. Two-thirds of the way through the contract, the consulting firm complained that the supplier was not expending enough manhours on inspection. Taken by surprise, the supplier checked his procedures and the quality of work done and found them to be to his satisfaction. Nevertheless, the consultant continued to complain, so the supplier hired a pair of independent inspectors to survey the quality of work done. These, too, confirmed that everything had been proceeding normally, so the supplier called for a meeting with the consultant. Arriving with a team of two, the supplier and fabricator found themselves at a meeting with the following personnel from the consultant: project manager, design manager, project coordinator, contracts manager, welding inspector, painting inspector, refractories inspector, scheduler, and refractories contractor, all charging to the job!

Immediately, the project manager went on the offensive, accusing the supplier of undermanning the project, using a list of minor complaints about the quality of the work, and referring constantly to threats of loss of schedule in making suitable corrections. Calmly, the supplier agreed to have the corrections made and the fabricator agreed to work extra shifts to protect the schedule. When the consultant continued to press his charges of undermanning the project, however, the supplier suggested that both the consultant and the supplier meet with his client to review their respective use of manhours. The project manager brought the meeting to a hasty close, and the matter was never brought up again.

Manhour padding is not always as blatant as this. However, one should realize that even with good consulting firms there are pressures on the project manager from within his firm to utilize available personnel. But there are also pressures from his client to stay on budget, or better yet, to *minimize* the use of manhours. No doubt the project manager has to walk a thin line, balancing one against the other.

Bad consultants are also adept at not taking responsibility. You can see it in the specifications and contract documents where all possible avenues for errors, omissions, or plain bad luck are covered by words intended to absolve the consultant of guilt and pass responsibility to suppliers, contractors, and even the client.

One consultant got called in by a client who complained that foundations were cracking on a row of fans. A young engineer who had been assigned by the consultant to trouble-shoot discovered that the designer had not followed the recommendations of the manufacturer and had elected to use a cheaper design. The young engineer was quickly removed and replaced by a more seasoned chap who found, by examining the maintenance record, that a blade of one fan had been replaced on warranty. On pressing the operators, he learned that the original blade had lost a piece of its tip and that a vibration switch had shut down the fan. By making an issue of it with the client, he was able to embarrass the fan manufacturer into providing part of the cost for foundation repairs. The client put up the money for the balance, and the consultant was able to make a small profit for redesign. Nobody thought to ask why all of the foundations had cracked, not just the one under the fan that had shed part of a blade.

Another consulting firm discovered that it had placed the counterweight for a belt conveyor in such a position that, if the belt broke, the counterweight would crash through several floors of an operating area, possibly killing workers. The correct solution would have been to relocate the counterweight. However, it would have embarrassed the consultant to admit his oversight. Instead, elaborate ways to catch a falling counterweight were studied—at the client's expense—and the faulty design was covered up. Engineers are human and therefore will make mistakes. Correcting them when they are discovered is what estimate contingencies are for, and in no way should such cover-up practices be tolerated.

Hero's Haven — What Good Consultants Can Do for You

If that last dissertation left a bad taste in your mouth, here's the good news: Most consulting firms depend on their reputations to sustain business and will not play the dirty tricks previously outlined, individual employees that come and go excepted. Indeed, if you have chosen a reliable company and their project manager shows these tendencies, they will be quick to remove him at your slightest suggestion. A good consultant can be the best investment you make in realizing your project, especially if the project has a moderate to high degree of complexity, creativity, or sophistication to it. If you have chosen well, your consultant can bring a broadening influence to your own project team and add a wealth of experience from similar competitive projects.

For most operating companies, projects of significant size are a sometimes thing; for consulting companies, they are a way of life. It follows, therefore, that using a good consultant will provide the smoothest line from conceptual design to start up.

For one thing, it is reasonable to expect to find repeat teams in a reputable firm. If you can use a group who have a track record of success on similar projects, you can expect to get at least as good results if not better. And if the company has depth, it will be able to reach into its well of talent to bail you out of trouble, should that be necessary.

Something that a competent group can provide is proven systems, be they computerized or manual. Usually both are available at your request and are flexible enough to adapt to your personal needs.

A good consulting firm will keep you advised on a regular basis, will give you the straight news—good or bad—and will do their utmost to avoid surprises. When things go wrong, whatever the cause, they will leap to the task of correcting the problem, putting your needs at the forefront.

One of the most sensitive, yet necessary, functions of a first-class consultant is to inject a sound measure of realism into plans for a project. It is almost universally expected that a client will approach a consultant with a slight-to-excessive degree of optimism. For example, it can usually be predicted that clients will underestimate the capital costs of their own projects. The role of the consultant is to ensure that expectations are realistic *before* embarking on major capital expenditures. It is not unknown for a consulting firm to kill a project despite the prospect of losing the work attached. A bad consultant may jump through hoops trying to make an unfeasible job feasible in order to get the work, but if he is true to his craft, he will bear the sad news when necessary.

The other side of the ledger is the most palatable. That is, a marginal situation can be made a success through the expertise of a knowledgeable consulting group. Good consultants can do that more frequently than is generally known.

CRITERIA FOR CHOOSING A CONSULTANT

Referring to Chapter 4 — When to Use a Consultant — it is assumed for this exercise that you:

- have estimated the size of staff needed to do the job and its cost;
- feel that you need more than a drafting service, that you need expertise.

It should be mentioned here that Chapter 2 — The Science of Project Management — should be reviewed, for the criteria that you would apply to your own organization also apply to a consulting firm. That is, you need

- the right people
- the right organization
- the right systems

These basics must be satisfied before you can expect success.

When selecting consultants, it is a good plan to have a prequalification round of proposals, especially if you are not familiar with which companies are particularly suited to your industry. At this stage, you can give consultants the opportunity to make their pitch to you without having to spend a great deal of money in preparing detailed proposals, a particularly frustrating and costly exercise that should not be done if in the end there is no chance for the bidder to get the contract. In addition, it gives you a chance to narrow the field to no more than five bidders and no less than three, thus reducing your workload when you prepare detailed evaluations. If you know the field well, of course, this stage can be eliminated.

Requests for prequalification proposals can be made in a simple letter, which should contain at least the following:

- – one paragraph description of your project, including its production capacity, estimated capital cost, and schedule;
- – a brief description of the scope of work for the bidder, for example, "total responsibility for engineering and procurement";
- – a request for proof that the consultant is qualified to perform the work and is able to bid.

That last, the ability to bid, should be emphasized. There is no point in reviewing a consultant's qualifications if he has no intention of bidding due to his current workload or for other reasons.

If you have requested a large number of consultants to submit qualifications packages, particularly if you do not have a firm idea of what each company is best at, you will be surprised by the responses. Some companies will simply decline because they are unqualified. Some will have no expertise whatsoever in your field, yet will try to convince you that they can handle the job. Others will be aware of their weaknesses and offer a joint venture or subcontract with another firm. And finally, word will get around and there will be a small flood of companies calling you and wanting to get in on the action.

Joint ventures should be avoided because at least one of the partners is in strange territory, and your project could suffer as a result. In addition, the likelihood of recruiting a team of personnel that has worked together before is diminished. If you feel that you must put two or more companies together to do the job, then it is better to insist that one company is subcontracted to the other rather than to make it a joint venture. This minimizes the possibility of a power struggle within the project organization and avoids the attendant delays.

Ideally, you should consider companies who have a *recent* history of doing similar work. If similar projects have not been done in the past ten to fifteen years, then you can expect that most of the key personnel involved would not be available.

Once your selection of qualified candidates has been made, review Chapter 12 — Specifications and Contracts — in preparing a request for bids. Be very specific when describing what the consultant is expected to do and what he is not expected to do (scope of work).

Following is a check list of most of the data that you need in your evaluation in addition to information about normal general conditions. This list can be included in your form of proposal:

– Type of company: private, partnership, corporation
– Last annual report
– Last financial statement
– List of officers and directors
– Signing officers for contract
– Past and pending litigation
– Principal type of business
– Length of time in business with current owners
– List of recent similar projects with dates and addresses
– Capital costs of these projects, giving original budget estimates, last budget estimates, variances
– Consulting costs of these projects, giving responsibilities, original budget estimates, last budget estimates, and variances
– Original and final durations of these projects
– Present status of these projects
– List of references who can be contacted regarding these projects
– Organization chart proposed with names of all supervisors given
– List of current projects and their status
– List of current projects pending
– Total number of people on staff
– Total number of people pending on other proposals
– Total number of people available for this project
– Amount of extra hiring, if any, required for this project; categories in which extra hiring will be necessary
– Manpower loading diagram
– Descriptions of systems proposed
– Sample documentation to be used
– Total office area
– Office area available for this project
– Space available for owner's personnel

If the project responsibility is expected to cover construction management, then the content of this list will expand to include a separate section for data on site facilities and equipment. If a complete engineer-procure-construct (EPC) program is planned, then the format of your form of proposal will be quite detailed, incorporating data included in Chapters 12 and 13, which cover specifications and contracts, and construction management respectively.

Working with a consultant requires a thorough knowledge of the project management process. The next eleven chapters are dedicated to setting down methods that should be used in project management. They are followed by a chapter on how to work with a consultant.

CHAPTER

6

PLANT DESIGN

A major seawater magnesite plant had been planned for some time. Process design, based on pilot testing and laboratory work, had been contracted. Major equipment sizing had been completed. At that point, the owner of the proposed plant engaged the services of a consulting engineering company to do the work.

After some development work and a visit to the site, the time to begin layouts arrived. At that point, the project manager called in his plant design engineer.

"I'd like you to have the model shop make up the major pieces of the plant to scale so that we can begin the layout," he announced.

"Oh, I don't think we need to do a model," said the engineer. "It's a fairly simple plant to design. Most of the equipment will be laid out on one level and there's not much in the way of process piping. Just leave it with me, and I'll have a preliminary layout done in a couple of weeks."

Pursing his lips, the project manager thought for a moment. "I really would like that model made. Then we can take pictures of it and give them to the client."

"You think that will impress him?"

"It might."

"Well," said the design engineer, "it won't help me any, but if you insist, I'll make up a model. In fact, the model components should be ready about the same time that my layouts are finished."

"You don't understand," said the project manager. "I don't want you to do *any* layouts. Just make the model parts."

Now the design engineer looked confused. "I know I can use a model to do design. I often do when there's a difficult three-dimensional interrelation between components or when there's a complex situation that could lead to interferences, but this. . . ."

"That's not the point," said the project manager, cutting him off. "What I want to do is bring in all the group leaders, all the disciplines, including construction. Then, we'll all gather around the table and design the plant — together."

"I see," said the design engineer. "In that case, you can count me out!"

"What?" The project manager was not accustomed to being contradicted.

"I'm sorry, but, as the saying goes, 'A camel is a horse designed by a committee.' "

"Nonsense."

"No sir. It's true. Although input to the design by everyone will have been theoretically made, in the end the most dominant parties will get what they want, and that may not necessarily be the best plant design."

"How do you know?"

"I'm a professional. I've seen it happen before."

At that point the project manager felt tempted to simply exclude the designer from the group and proceed, or else replace him with a more cooperative fellow. But his track record in design was excellent.

"How can we get input from everyone who is important, doing it your way?"

"Let me do the design. They can look at it and criticize. I'm not inflexible, I can make changes, if necessary. Like I said, I'm a professional."

Reluctantly, the project manager agreed, and a successful design resulted.

Plant designers who are true professionals are an extremely rare breed, and most of them complain about being pressured by people "who don't know what they're talking about." That statement, in itself, speaks for the lack of esteem in which the professional designer is held. Although there are, perhaps, a hundred plants of various types being designed in North America at this writing, and possibly double that number around the world, only a handful will be done by specialists in plant design. The rest will be done by committees and a variety of technical people who are skilled in their own fields but who have been parachuted into a jungle with which they are not entirely familiar. That is not to say that these plants will not be successful. Indeed, the vast majority will earn their owners handsome profits. A professional layout man may tour them, however, and pronounce that 80 percent of them are badly designed. Why?

To begin with, plant design has never been addressed by educational institutions or industry itself as a distinct skill, possibly because there are a limited number of plants built in a specific company's history. The progression from acceptable to excellent can only have a limited number of steps. Even then, if a bad plant design is considered successful in a given industry, who is going to do the next design? Same group, right? Bad design practices get perpetuated. Indeed, one sees the same mistakes made over and over. The saving grace is that they still work: They still make money for their owners.

If you wish to see a good design done, make sure you put a professional in place, and back him up with people who have skills in process engineering, construction, operating, and maintenance. His job will be to orchestrate their ideas and mould them into a compromise that is acceptable to all of them, because a good design is one that optimizes many different aspects.

INPUT FROM PROCESS, OPERATIONS, AND MAINTENANCE

If you let a process man design your plant, chances are it will be difficult to operate and even more difficult to maintain. If you let an operations man design it, the equipment will be grossly overdesigned and expensive. If you let a maintenance man design it, access to equipment and provision for stand-bys will be paramount, and the shops will be a showcase of automated machinery and tools. But if you do not involve each of these men or their departments in the design, it may simply not work well at all. It is equally important that you involve them as a team, for each will have to compromise his ideals for the good of the others and the general economic health of the plant.

Process engineers often have a fixation on the accuracy of stream flows and their control to the exclusion of the practicality of the machinery that makes it work. Little attention is given to upset conditions in many cases, so that from a materials-handling point of view things can get difficult. When a process stream breaks down, it becomes necessary to handle what has backed up before it as well as to dump or divert the stream itself. Operators try to have this flexibility built in; process engineers think you do not need it.

What an operator is looking for is a smooth-running plant that requires no manual monitoring and that has large tolerances in process criteria so that little attention is needed to control it. In addition, he wants machinery that will not break down.

A maintenance man requires operators who will be gentle and a process that will not corrode or otherwise harm equipment.

The plant designer has to satisfy these requirements from a layout point of view, but the project manager has to see that everything can be done within budget. Periodic reviews by all parties must bring these varied concerns together in a rational compromise.

Repeating Old Errors

The frequency of repeated errors made in designing industrial plants is amazing. The following is a typical list:

NO PLANS FOR EXPANSION Even if it appears that your plant will never increase in size or capacity, sometimes it still happens, or sometimes new product lines are added. Therefore, space should be allowed for this possibility, either within the building structures or within the real estate boundaries.

USE OF TALL, MULTISTORY STRUCTURES Often a tall process plant is used when ample real estate is available to spread it out. This may be dictated by process but is usually avoidable. The following disadvantages arise in multistory structures:

- Power costs in process materials handling are higher. It is cheaper to move materials horizontally than vertically.
- Heating distribution is more difficult due to migration of heat from lower levels to upper levels.
- Support steel becomes more massive with height.
- Vibrating machinery at upper levels requires mass damping, which can be more costly than floor-mounting in concrete floors.
- Maintenance and operating personnel need to do more climbing, which is more tiring, even with the help of elevators.
- Lifting equipment and materials to upper levels is more difficult and costly.
- Visual contact between areas is not possible without grating floors, and then it is limited.
- Things fall from one level to another, increasing risks and contributing to untidiness.

USE OF HIGHLY IRREGULAR BUILDINGS   Often a building is designed with a myriad of lean-tos, penthouses, jogs, els, and niches. Sometimes this is done to reduce the area of siding and is justified on a capital cost basis. What happens in practice is that the number of intersections increases the number of weather joints, which are potential leaks and potential maintenance problems. In parts of the world that have moderate to heavy snowfalls, snow pockets are created. These require heavier structures to support them and the deflections created by their weight can lead to further leakage problems.

THE CATHEDRAL EFFECT   An antithesis to the paragraph above, this effect is achieved by nonstepped rooflines. It results in massive volumes of unused space that often need to be heated or cooled.

ORIENTATION WITH PREVAILING WINDS   Chemicals, dust, moisture, and heat are frequently given off upwind of plants, creating personnel and sometimes operating problems. Transformers and insulators downwind are subject to shorting out. Short-circuiting of wastes to air intakes occurs many times when local winds are not taken into account.

ANGLED INTERSECTIONS   In many instances, galleries and other junctures are made at other angles than right angles. This invites construction errors. It also can create internal problems when hanging equipment (in the case of sloped walls). Belt conveyors are difficult to load properly at an angle and can create belt-tracking problems and promote spillage if they are not loaded inline or at right angles.

TOO MANY TRANSFER POINTS   Where materials-handling equipment, such as belt conveyors, is used, it is important to avoid transfer points because they create the potential for spillage and increase maintenance. Often, careful layout can eliminate many transfer points.

LOCATING EQUIPMENT IN PITS   Pits in the floors of plants attract garbage and become receptacles for liquid spills. In addition, it is frequently difficult to maintain any equipment located in them, and they are difficult to clean. The most common error made in this regard is the location of conveyor tail pulleys in a pit in the floor. Justification for this is usually based on overall building height, but careful layout can help avoid pits.

POOR CLEAN-UP CONDITIONS   Access for floor clean-up is usually neglected. Space should be left beneath equipment to sweep or shovel up dust and spillage. When washdown is anticipated, floors should slope to sumps, which in themselves should have a gradual slope on one side for shovelling out residue or solids. Grating floors should not be used if material will fall through them to the floors below.

VIBRATING EQUIPMENT IN POOR LOCATIONS   Vibrating equipment, such as screens, should be located on the ground floor when possible. If located on upper floors, avoid midspan locations. It is surprising how often natural frequencies of building structures are not checked, resulting in the whole building shaking when equipment vibrates.

COORDINATION OF LAYOUT OF CABLE TRAYS   It is not uncommon to see cable trays transporting liquid spills.

UNPRESSURIZED ELECTRICAL ROOMS   Even in plants with fairly clean atmospheres, electrical, instrumentation, and computer rooms are frequently plagued by dust. Pressurizing them slightly from the ventilating, cooling, or heating outlets will minimize this problem.

LACK OF HEADROOM   Building codes generally specify headroom requirements for personnel, but invariably pipes, cable trays, and substructure beams reduce available headroom. Anticipation is the key.

POOR ACCESS TO EQUIPMENT   Maintenance people constantly complain about lack of access to equipment for repairs, and in a surprising number of cases, replacement is impossible without dismantling buildings or substructures.

CONSERVATISM IN DESIGN

Making judgements on equipment selection and sizing will affect plant availability. Obviously, you can overdesign to the extent that availability approaches 100 percent, but at what cost? During the preliminary design stage, a calculation can be made by bringing together the process, operating, and maintenance personnel. By analyzing potential downtimes of each piece of equipment in the process stream and doing a statistical probability analysis, the overall availability can be estimated for the plant. From the results of this calculation, the throughput can be determined on an hourly or other unit basis. At that point, a judgement must be made, moderated by cost, as to how much conservatism is desired.

Naturally, the operators will push for as much extra capacity as they can get so that they can crank up production whenever they get behind schedule. Maintenance engineers may provide some support for this idea, but those who are experienced will realize that operators tend to push equipment to its limit. After all, their performance is recognized on a production basis. If they can keep the coffers full, they will win brownie points. The maintenance men will want extra stand-by equipment and easy means to switch over as a priority. You will want your process people to make sure both are compatible with the qualitative aims of production.

In many industries, conservatism in design is inadequate, and a rag-tag operation results. Other industries tend to become more conservative with time, building on experience. A danger there is when ultraconservatism replaces the quest for optimization, and you wind up using a sledge hammer to drive a tack.

Astute professionals will realize this and try to use their influence to control the design. A lot of judgement is needed, even at the detail stage. For example, a lifting lug may be designed to be made out of one-quarter-inch-thick steel plate, and a calculation may show that it has a safety factor of six, based on load. A good designer may recognize that atmospheric corrosion in the area could be severe and that, as a result, mechanics may tend to drive the pins in with a heavy hammer. Sometimes they miss and hit the lug. A lug, weakened by corrosion and with stress concentrations due to nicks from a hammer, could fail. He decides, therefore, to make it one-half inch thick. Invariably, that kind of design gets copied in the next plant, where that same condition exists. Someone there gets nervous about another condition, and the lug becomes five-eighths of an inch thick. And so on. Unchecked, that progression can produce very expensive plants indeed.

USE/MISUSE OF MODELS

Making models of your plant as a design aid became popular a couple of decades ago. Unfortunately, many project managers underestimated the cost of producing one of these and cavalierly had them built as an adjunct to the design office. Model makers took advantage of them, of course, and a lot of miniature plants were built which looked very pretty but were of little use and invariably put quite a dent in the budget.

Models are an extremely useful method of checking out or solving complex process plant design problems, but they are expensive and their use should be carefully planned. Some industries, notably refinery and chemical, use models for actual design, making drawings from the model instead of the other way around. A model of a strip mill, on the other hand, serves relatively little purpose. In general, when multidiscipline layouts need coordination, such as piping, ducting, and cable-tray running, then a model can pay for itself in field cost savings by enabling interferences to be identified in the office and not at the site.

A model of your plant can become a permanent tool for training personnel to operate and maintain it. A much quicker and more accurate insight into where things are and how they interrelate can be given by using the model in the office, and it sure beats climbing and walking and explaining things in a noisy industrial atmosphere.

DESIGN AUDITS/CHECKING

Earlier we spoke of design audits from the point of view of ensuring that engineering calculations were correct (page 23). Probably some of the most costly items in the field, however, are prefabricated steel, equipment, and piping that do not fit. Yet often, dimensional checking in the design office is not done. Engineering offices must do dimensional checks on a routine basis. Normally, checkers go over each dimension on the drawings with a yellow and a red pencil. Good dimensions are blessed with yellow, red is reserved for flagging errors. Once the corrections are made on the original, a green mark is made on the print, and it is filed as a record print. The inspector's function is to see that the equipment and the prefabrications conform, operationally *and* dimensionally, to the drawings prior to shipment to the field.

CHAPTER

7

PLANNING AND SCHEDULING

BAR CHARTS

Everyone who has been involved in a project understands the use of bar charts in planning and scheduling, and that, precisely, is why they are currently used and will continue to be used. Figure 7.1 is a typical project schedule using a bar chart. Figure 7.2 shows an update.

PERT AND CPM NETWORK ANALYSIS

Project Evaluation and Review Technique (PERT) was first developed for the United States Navy for large projects and was initially used on its Polaris weapon system in 1958. Today most government programs in the United States use this system of planning and scheduling. Booz, Allen & Hamilton, a consulting firm, are credited with aiding the navy in its development.

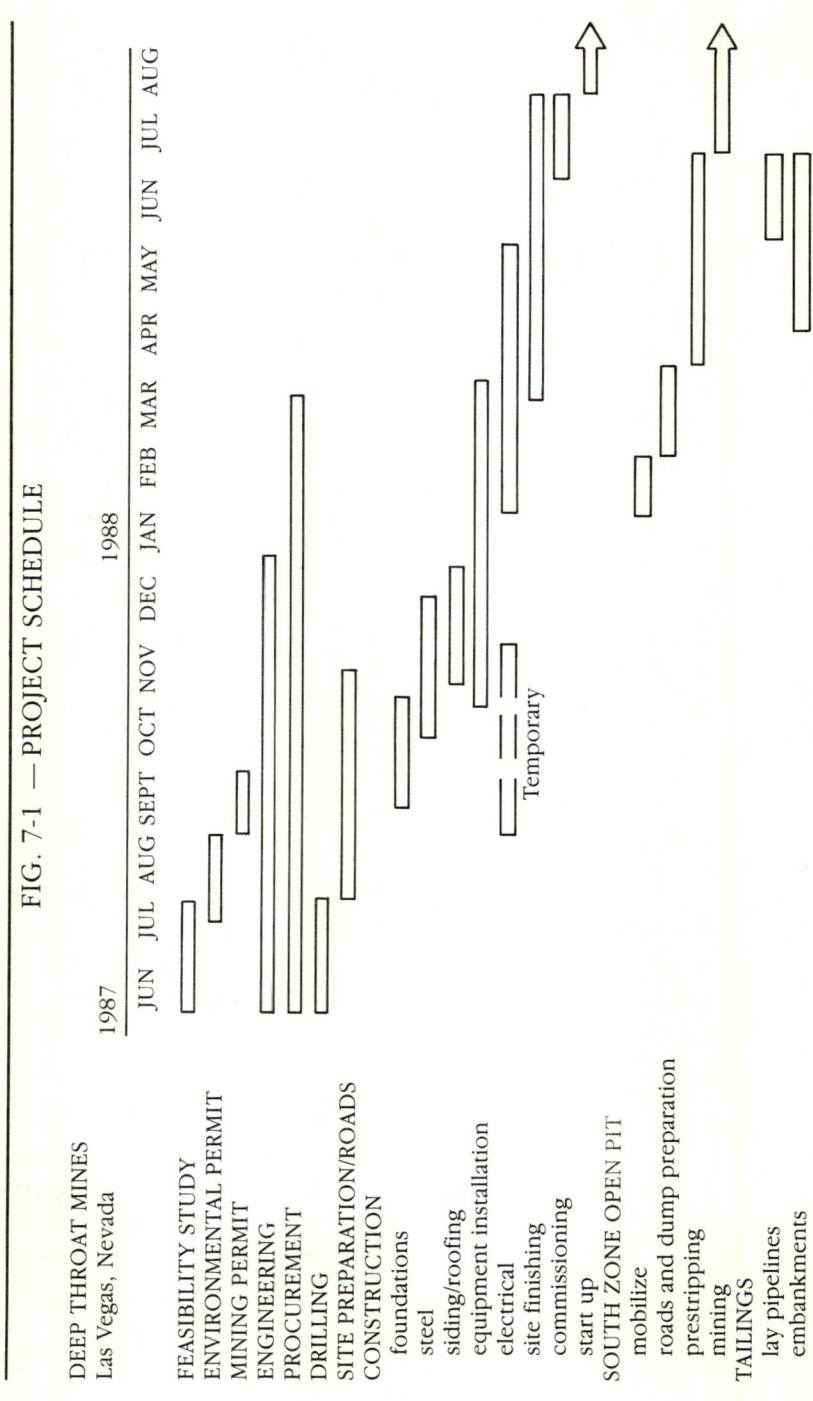

FIG. 7-1 — PROJECT SCHEDULE

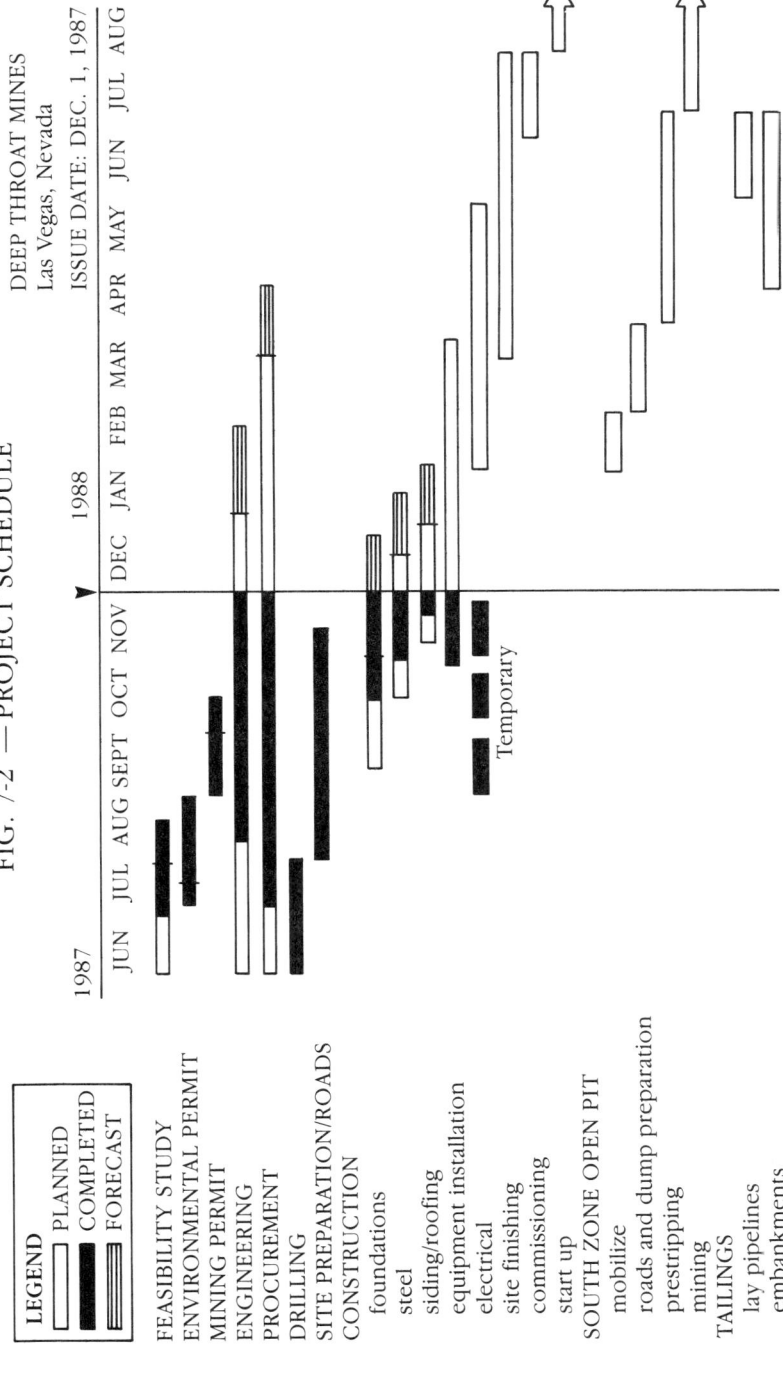

FIG. 7-2 — PROJECT SCHEDULE

DEEP THROAT MINES
Las Vegas, Nevada
ISSUE DATE: DEC. 1, 1987

The Critical Path Method (CPM), which is somewhat similar to PERT, was developed by the Du Pont Company for controlling construction of large chemical plants. It is probably the most common method for planning and scheduling in use today in private industry.

PERT

In the early 1960s, PERT was used mainly for scheduling but later was expanded to include the relationship between cost and time, which were designated as PERT/time and PERT/cost.

A PERT/time network comprises events and activities. An event is the result of one or more activities that take time. In building a shed, for example, completion of the foundation is an event. The activity of building it takes time. Represented on a network, building a foundation would appear as follows:

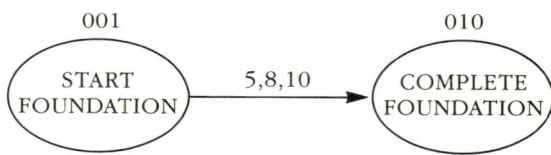

Event 001, starting the foundation, leads to completion, event 010. The most optimistic time to do the job is five hours; the most likely, eight hours; and the most pessimistic elapsed time is ten hours. Thus, the activity is represented by five, eight, ten.

Figure 7.3 shows a PERT/time network for building a shed. Obviously, more than one man is working on the shed in our example, so that trusswork, framing, and foundations can be constructed simultaneously.

A critical path and slack times are computed as follows:

First, an averaging formula is applied to reduce to a single expected time (TE) the three time estimates for each activity.

$$T_E = \frac{a + 4m + b}{6}$$

a = optimistic
m = most likely
b = pessimistic

These single times are shown in Figure 7.3 as bracketed numbers. In addition, standard deviations (δ) and variances (δ^2) can be calculated from

FIG. 7-3 — PERT/TIME NETWORK FOR BUILDING A SHED

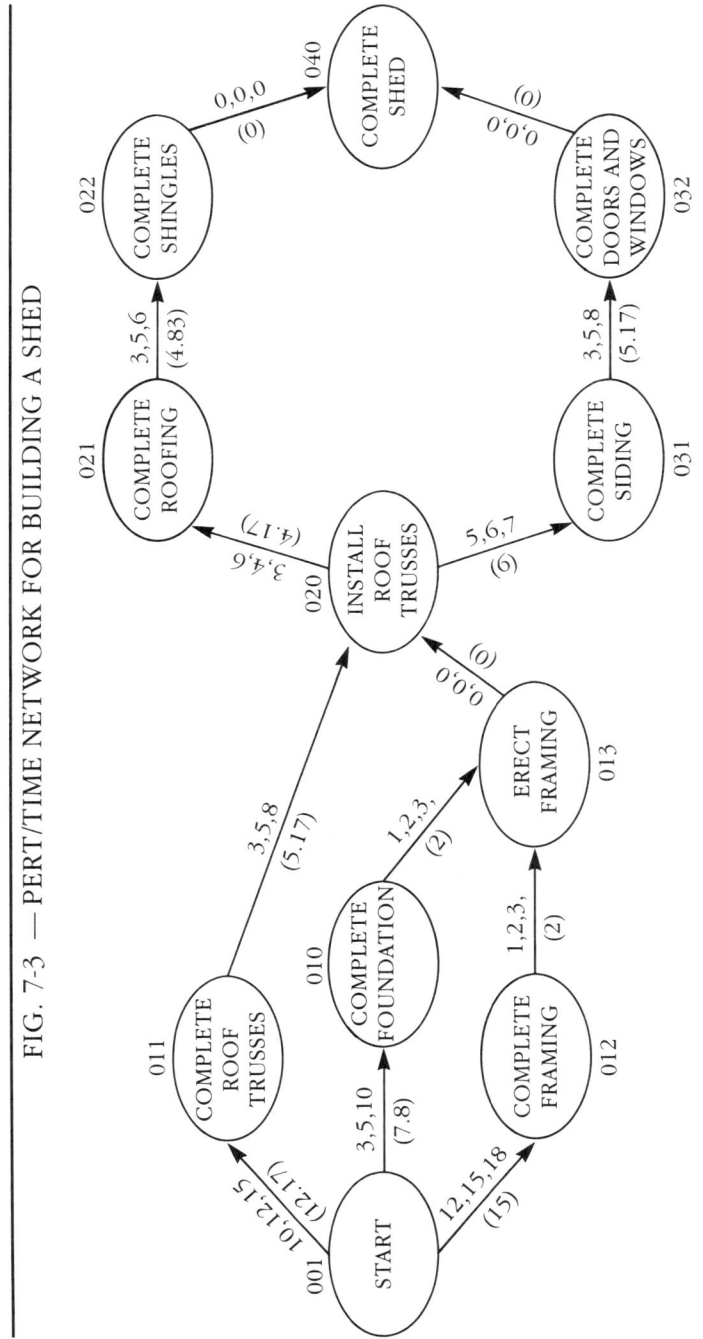

$$\delta = \frac{b - a}{6}$$

The critical path is the path that is taken that will require the greatest expected time to accomplish. In Figure 7.3, it runs as follows:

Event	Time
START	(hours)
Complete roof trusses	12.17
Install roof trusses	5.17
Complete siding	6
Complete doors and windows	5.17
Complete shed	28.51

To compute the slack time, the process is reversed, starting with the final event. A slack order report (Figure 7.4) is made as follows, where the "expected elapsed time" (T_E) and the "latest allowable time" (T_L) for each event are listed. Note that on the critical path $T_E = T_L$. Computation of variance and standard deviation is optional and involves adding the variances for each activity along the critical path.

FIG. 7-4 — SLACK ORDER REPORT FOR BUILDING A SHED

Event	T_E	T_L	$T_L - T_E$
001	0	0	0
011	12.17	12.17	0
020	17.34	17.34	0
031	23.34	23.34	0
032	28.51	28.51	0
040	28.51	28.51	0
022	26.34	28.51	2.17
021	21.51	23.68	2.17
013	17.0	17.34	0.34
012	15.0	17.34	2.34
013	9.8	17.34	7.54
010	7.8	17.34	9.54

When using a PERT/cost network, usually only T_E values are shown. Activities are given cost values based on manpower requirements. A calculation using this system sometimes highlights a manpower problem that can be solved by adjusting slack times in a manpower-levelling exercise.

CPM

Networks similar to PERT are used in CPM. The essential difference is that CPM uses "early start dates" and "latest finish dates" instead of "optimistic," "most likely," and "pessimistic" activity times. "Float" is similar to "slack time."

Total float can be calculated in a similar fashion to that used for calculating slack time in PERT. Referring to Figure 7.5, the following calculation applies:

Event number	Duration (days)	Early start	Early finish	Late start	Late finish	Total float, days
001	0	0	0	0	0	0
010	8	0	8	8	16	8
011	12	0	17	0	17	0
012	14	0	14	2	16	2
013	1	15	16	16	17	1
020	5	12	17	12	17	0
021	4	17	21	19	23	2
031	6	17	23	17	23	0
022	5	21	26	23	28	2
032	5	23	28	23	28	0
040	0	23	28	23	28	0

Starting with a forward pass, early start and finish times are calculated. Then a backward pass determines the late start and finish times. Total float can be calculated as the difference between early and late start times. Events having zero float are on the critical path.

CPM, like PERT, is sometimes used in conjunction with a cost evaluation. On the assumption that a crash schedule for each activity would cost more, a dual evaluation is made using different durations in order to determine if a tight schedule has any economic advantages. Sometimes, by increasing the cost of some of the items that are on the critical path, a net benefit results.

COMPUTERS

Both PERT and CPM calculations are commonly done using computer software. Programs are available that print out detailed

FIG. 7-5 — CPM NETWORK DIAGRAM FOR BUILDING A SHED

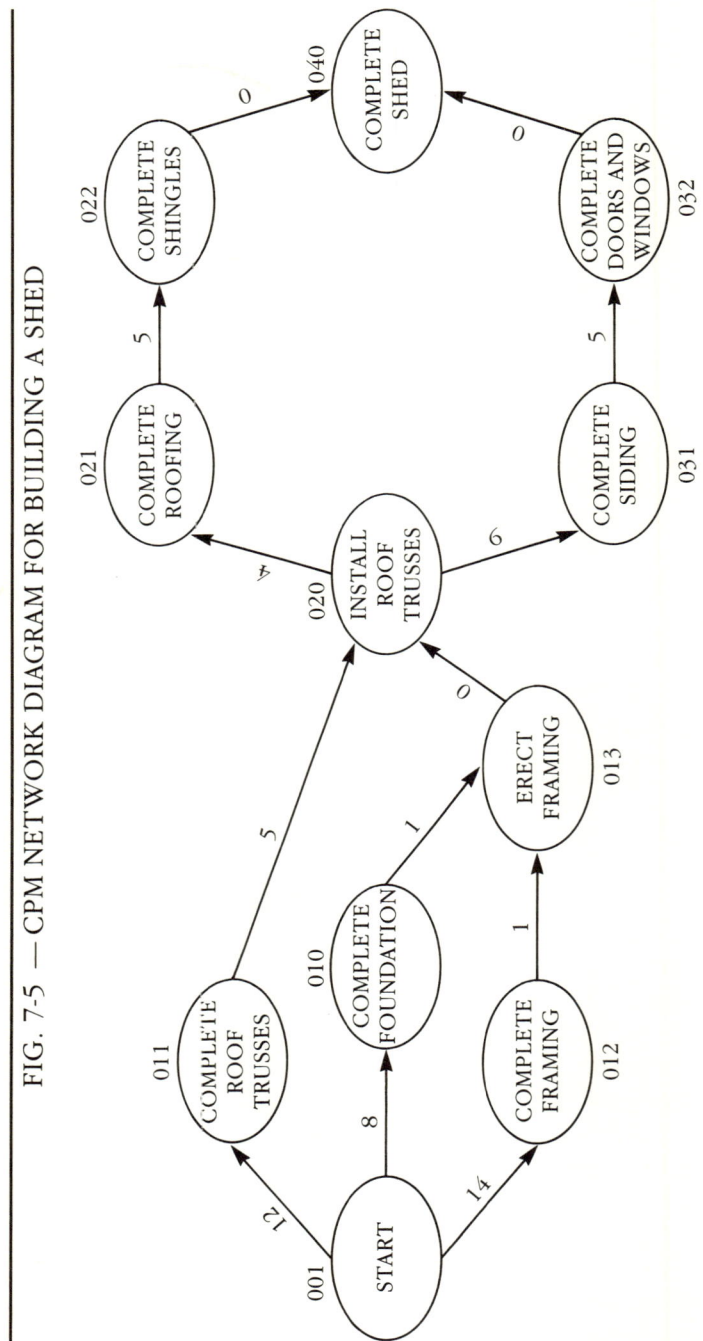

networks, distill them into major overall milestone networks, and also convert them to bar charts.

FLOAT: A DOUBLE-EDGED SWORD

Once a critical path is determined and a schedule is published for all to see, a CPM similar to that shown in Figure 7.6 is produced, and this becomes the operating document for managing the project.

Note that float can be used to suit manpower availability. For example, there are two hours float available to erect the framing, activities 010 and 012 to 013. This float could be planned for activities 001–010 and 001–012 on the assumption that the longer the schedule time, the greater the odds are that it will overrun. A similar flexibility could be applied to activities 020–021 and 021–022.

Obviously, an ambitious project manager will attempt to use his float wisely, not spending any of it if at all possible. By publishing a CPM schedule and showing all of the available float, however, he puts himself at a psychological disadvantage. What incentive has the foundation builder to complete the foundation in our example in eight hours? Chances are he will dribble away his float and then some, and your framing erection labor will be pushed against the wall. Parkinson's law will prevail:

Work will expand to fill the time available.

The determination of float in calculating a CPM schedule should therefore be regarded mostly as a strategic manpower-levelling tool, otherwise it will merely tend to sabotage your productivity.

REALITY

Earlier, we talked of a metallurgical project that ostensibly failed due to placement of the wrong personnel. Subsequently, the consulting engineering company got the bum's rush. Just prior to that fateful event, however, there was a distinct odor of defeat about the project. A trouble-shooter was dispatched to the project office to see if anything could be salvaged. What he found was a worried project manager peering intently at a 5000-activity CPM computer printout tacked around four walls of the boardroom. Our trouble-shooter knew right then that all hope of saving the project had evaporated.

FIG. 7-6 — CPM SCHEDULE FOR BUILDING A SHED

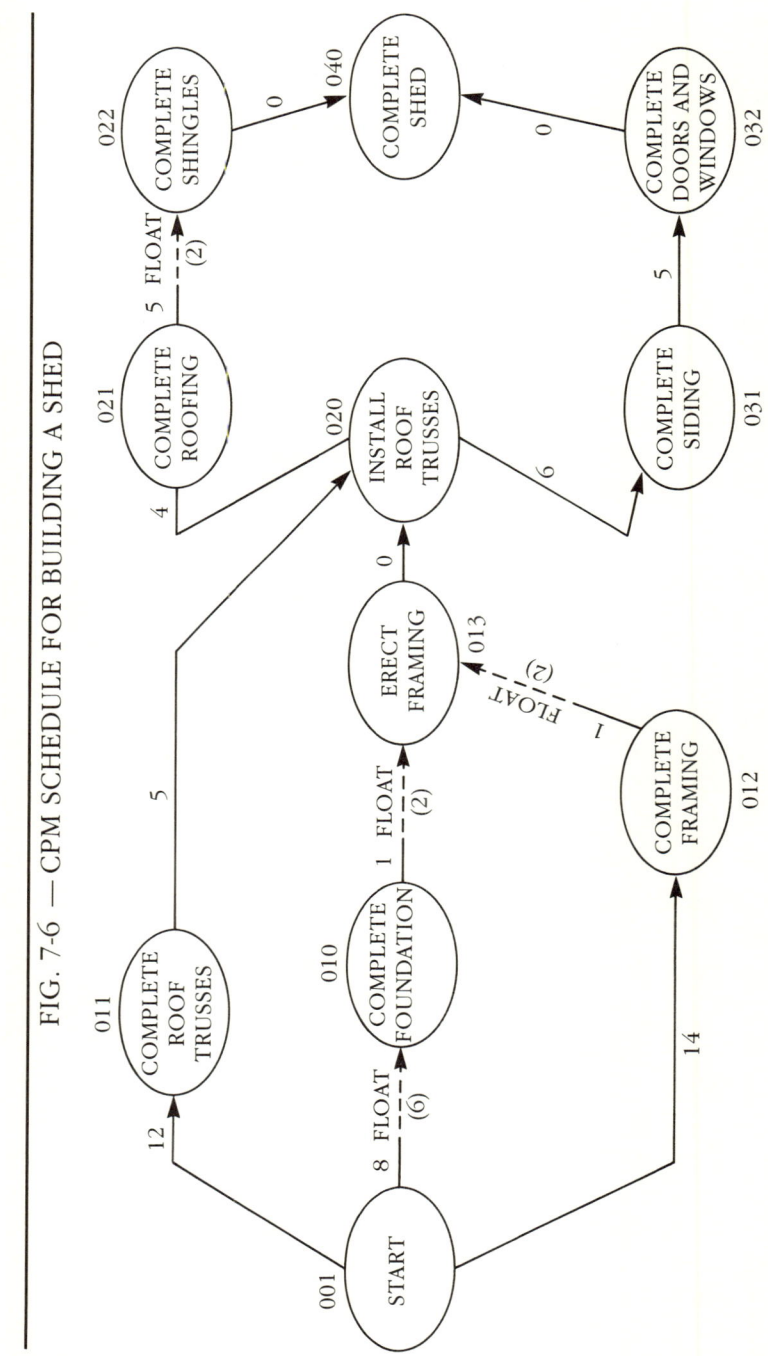

Another large project appeared to be in trouble. Nobody seemed to know exactly where the project stood; it had fallen into the "two-thirds malaise" that is discussed later in this book. A panel review was called, and its chairman, a seasoned project manager, asked for a review of the schedule. At that point, the chief planner reddened and said, "The current schedule is out of date."

"Then update it."

"That," said the planner, "will take six weeks."

"Why? Don't you have it on computer?"

"Well—er—yessir, but getting all of the input data will take six weeks. After all, it's a 13,000-activity CPM!"

The chairman rolled his eyes, then looked off into space for a moment. "And then, I suppose, it will be six weeks out of date?"

"Oh, I wouldn't want to speculate. The computer could be broken down or otherwise not available."

PERT and CPM suffer from the tendency of users to expect too much from them. It is always tempting, when a computer is available to do your work for you, to overcomplicate to the point of utter confusion.

In preparing a work breakdown, which is the essence of planning and scheduling, a great deal of professional judgement must be used to distinguish between what is controllable and what is not and to decide to what level of detail you should go. If we take our simple shed project, for example, thirteen activities could become a hundred or more, simply by going to a more detailed breakdown of activities. The steps going from 001 to 011, completion of trusses, could be subdivided into the following activities:

> purchase of nails
> purchase of lumber
> cutting of gussets
> cutting of members
> building a jig
> nailing the gussets

If the approach to the project used the methods of George Gillycuddle, i.e., made a meal of the job, and if a serious attempt was made to monitor all of the activities, then control of the project would take more time than doing the actual work!

Intelligent use of CPM or PERT requires the confidence that your subnetworks are realistically done, i.e., that the right people are doing them. (Garbage in/garbage out does no amount of good!)

The project manager described in our metallurgical plant example needed only to look at a schedule showing major milestones to see where he was going.

People in general have difficulty reading a CPM schedule, and so it is common to have the information translated into bar charts. This can be done manually or by computer. Float and critical path also can be transcribed.

Burning the CPM

Probably the greatest value that CPM has is that it forces all activities to be visualized by the project team in a clear enough manner to be put down into a network. This can do wonders for organizing a project and getting the team to pull in the same direction. As a day-to-day operating tool, however, it is grossly overrated. The need for converting to bar charts speaks for itself. So do the hundreds of overrun projects and a similar number that suffer from poor productivity, in spite of diligent use of the CPM.

Scrap the idea? Heck no! On the contrary, use CPM to plan and organize your project. Calculate the critical path with it. *But do not publish it!* Instead, convert to bar charts and use these as day-to-day operating tools. File the CPM in a drawer, lock it up for the record, and *burn all other copies!*

Why? First off, many team members will not be able to read it or will find it confusing at best. Secondly, float will be depicted, and, as project manager, you will want that to remain your little secret.

Be prepared to do other CPM runs at each estimating update or from time to time as project malaise threatens. Also, you must remember that the critical path can and will change with time and therefore will need identifying.

To recapitulate, the procedure for scheduling should be as follows:

- Prepare a work breakdown.
- Prepare a CPM network.
- Inspect and adjust the network for manpower levelling.
- Prepare a CPM schedule.
- Convert to bar charts for monitoring.
- File the CPM schedule and burn all extra copies.
- Recalculate as necessary.

Above all, make sure that input is subdivided into manageable portions and is made by qualified personnel.

CHAPTER

8

ESTIMATING

"You said *how much*?!!"

A truly professional estimator can hear such a wail without flinching. He is used to owners and project managers having overly optimistic ideas about a project cost. He has been listening to the project manager trying to keep the lid on his engineers, who want the best of everything. He has resisted the temptation to lowball his estimates in an effort to keep the project manager happy because he knows that it will come back on him later if a poor estimate is done. In fact, if he has done his work properly — and the project manager knows it — there is one and only one way to reduce the estimate and that is to change the scope of the work.

There is a temptation for engineers to do their own estimates. Many times an accurate enough estimate can be done. If there is a limited opportunity to go to the well for money, however, it is best to let a professional estimator have a go at it. It will gain you credibility to have an arms-length estimate, and you will be less likely to succumb to the temptation to push costs high or low.

Estimates may seem vulnerable to influences from economic and operating forces. Nevertheless, the results have generally been good over the years, with the exception of the period of rampant inflation in the 1970s. It is amazing how well a good estimator can predict the cost of a plant with very little information.

TYPES OF ESTIMATES

Three basic types of estimates are common in industry. In order of increasing accuracy, these are:

- Order-of-magnitude
- Preliminary
- Definitive

Occasionally a fourth kind of estimate is designated by some companies. Usually this is known as the update estimate.

Order-of-Magnitude Estimates

Sometimes known as a "ballpark estimate" or a "horseback estimate," the order-of-magnitude estimate provides a figure that is based on limited information, and, as a result, accuracies of only 20–40% can be expected. Usually an order-of-magnitude estimate is used for determining feasibility of a project. Some feasibility studies demand a higher degree of accuracy and hence a higher level of estimate.

Order-of-magnitude estimates are normally factored from historical information, such as:

- Cost per unit capacity
- Cost per connected horsepower
- Cost per area of land or building
- Cost of a similar plant, adjusted for site factors
- Sources and storage costs of raw materials

A limited amount of engineering is needed for an order-of-magnitude estimate. Normally, a preliminary process flowsheet, a plot plan, and building arrangements are required.

Preliminary Estimates

Preliminary estimates follow a feasibility study and are sometimes done to confirm the viability of a project. The estimate then can be used to obtain appropriation of funds and can serve as a preliminary budget for the project.

In order to prepare such a budget, enough engineering should be done to freeze the overall concept of the project and to define the scope and battery limits for detailed engineering and construction. The following serves as a minimum list of items for a preliminary estimate:

- Process flowsheet
- Plot plan
- General arrangement drawings
- Infrastructure definition
- Electrical single-line diagrams

Typically, preliminary estimates are made with a combination of cost quotations from suppliers, quantities, take offs, and factored data from similar projects. Overall expected accuracy for a preliminary estimate falls in the range of 10 to 20 percent.

Definitive Estimates

Sometimes a definitive estimate is looked upon as a major milestone with respect to committing the plant to construction. From that point of view, an estimate should be made to an accuracy of 5 to 15 percent and should be used as a firm control budget in building the plant.

Engineering should be better than 25 percent complete. In some cases, preliminary construction may have begun, and purchase commitments may have been made on long-delivery items. Firm designs and scope definitions should have been made in the following areas:

- Plot plan and building arrangements
- Flow diagram
- General arrangements of process plants
- Equipment specified as firm
- Foundations fixed in design
- Construction schedule

Definitive work should have been done on the following:

- Piping and instrumentation diagrams
- Electrical single-line diagrams
- Piping and electrical layouts
- Architectural requirements
- Infrastructure

On occasion, complete scope definition on a section of a project is not possible due to lack of information or for other reasons.

Such a section should be broken out of the project and estimated separately.

Major equipment should be quoted on a firm basis for a definitive estimate, with minor equipment costs defined by preliminary quotation. Quantity take offs should have been made on construction materials to the extent defined by engineering drawings. Labor pricing, productivity factors for the area, and subcontract prices should have been obtained. Costs for engineering of the balance of the work and for construction management should also have been firmed up.

Definitions

A typical industrial project estimate is made up of the following components:

> Direct costs
> + Indirect costs (sometimes called distributables)
> + Home office costs
> + Contingency
> + Escalation
> _____
> = Total cost

These components are defined as follows:

DIRECT COSTS These costs consist of material and labor used directly in the construction of the plant as well as in subcontracted work, but they do not cover costs to supervise or service construction labor. Direct costs are the total of the following:

– Direct materials, including plant equipment, permanent concrete, structural and building facilities, yardwork, roads, piping, electrical and instrumentation materials
– Direct labor, covering wages and burden for tradesmen, workers, and foremen who are involved in the installation of direct materials
– Direct subcontracts, covering complete materials and installation packages such as earthmoving, painting, and insulation

INDIRECT COSTS Some companies call indirect costs distributables because they apply to the overall project. For cost-coding purposes, they are often allocated to different areas of the project on a proportional basis.

Typical indirect costs consist of:

- Temporary facilities (camp, warehouse)
- Construction equipment, tools, and supplies
- Construction utilities (power, water, heat)
- Field office costs
- Insurance
- Payroll costs
- Construction supervision

HOME OFFICE COSTS Home office costs may include engineering, procurement, planning and scheduling, cost control, personnel, labor relations, legal fees, administration, accounting, and construction management.

CONTINGENCY Contingency of an estimate should not be confused with the accuracy of an estimate, even though an arbitrary subjective percentage is used to calculate contingency. It should be noted that *the amount of contingency is expected to be spent*, since, by definition, it covers costs that are regularly encountered but are impossible to define: errors, omissions, labor disputes, delivery shortages, or unusual weather.

Most contingencies used vary from 5 to 20 percent and may be applied to the overall project estimate or to individual components of the estimate. Generally, the less accurate the estimate, the higher the contingency.

ESCALATION Escalation may be calculated from government statistics on price rises expected between the date the estimate is prepared and the dates that actual costs are incurred. It can be estimated for the overall project or for major individual components, from which a weighted average is calculated.

ACCURACY

Accuracy of an estimate can be defined as the degree of confidence the estimator has in the prices he has obtained, the quantities he has taken off, and the allowances he has made. It will be backed up by a basis of estimate, which is a written description of his source, of the scope of work, and of assumptions made.

Usually accuracy is expressed as a plus-or-minus equal percentage such as ± 20 percent. One only has to take a brief glance at the history of projects to see the tongue-in-cheek attitude that is common in writing the minus sign. It is rare indeed to witness an underrun project, and when that happens, it is even rarer to see it reach the limit of the predicted accuracy. When submitting an estimate of 20 percent then, you will gain more credibility by saying "+20 percent to −5 percent" or even "+ 20 percent" period.

When estimates are submitted, the accuracy is often understated or buried in a lengthy written discussion for fear of scaring off the decision makers. It is unfair, however, to present an estimate without defining the maximum upset cost clearly so that the board or money appropriators can be forewarned of the possibility of having to dig deeper. A restatement of the components of an estimate should then be made as follows:

$$
\begin{array}{l}
\text{Direct costs} \\
+\ \text{Indirect costs} \\
+\ \text{Home office costs} \\
+\ \text{Escalation} \\
+\ \text{Contingency} \\
\hline
=\ \text{Total estimated cost} \\
\hline
+\ \text{Accuracy} \\
\hline
=\ \text{Maximum upset cost} \\
\hline
\end{array}
$$

Astute directors of a company will protect themselves against upset cost by reserving a source of funds to cover the accuracy. If several projects are in progress, then the sum of all of the estimate accuracies need not be protected, but instead, statistical analysis or a best guess as to how many of the projects could overrun and by how much should be made so that appropriate funding will be available if needed.

CHANGE OF SCOPE

A well put together estimate will be tied to a well-defined scope of the project in sufficient detail so that changes of scope are easily identified. Invariably, changes are made in the execution of a project that will affect costs. These should be identified if the estimate is to survive as an entity.

Project managers will cling to every change of scope that will affect the final cost of the project because it is their only protection against criticism if their projects are overrun. Most often, when a project threatens to overrun, things get a little testy, and the need for a definition of exactly what constitutes a change of scope arises. Which changes should be covered by contingency? Which changes by accuracy? It is important that these differences be clarified with all concerned *at the time of the estimate.*

Changes of scope result when the concepts originally envisioned by the project team — and the project users — are changed to entirely new concepts.

An example of an obvious change in scope is:

Original scope: Cinder transport from grate cooler to storage pile, using two belt conveyors

New scope: Cinder transport from grate cooler to storage pile, using one bucket elevator and one screw conveyor

Net change in cost: $50,000

Net change in schedule: No improvement

If, in the above example, the project team decided to stay with the original scope but to increase the width of the two belt conveyors from 24 to 30 inches, resulting in an increase in cost of $25,000, then an argument could ensue. Even though the estimate may have allowed for 24-inch wide conveyors, the increase in cost has to be absorbed by contingency because the engineers erred in selecting the narrower conveyor.

CODES OF ACCOUNTS

Costs of constructing a plant should be classified to allow development of investment records, but it is just as important to set a basic framework for controlling the cost. Usually codes of accounts are structured around a work breakdown by area or by sections on a flowsheet. For example, in a concentrating plant, major breakdowns could include: crushing plant, grinding area, flotation area, or drying area.

Some codes of accounts are regulated by law. For instance, the Federal Power Commission in the United States dictates that a strict code of accounts must be followed in the construction of power plants.

Cost coding begins and is structured at the estimating stage. Its primary use is in cost control and will be described in further detail in Chapter 9.

PREPARATION OF AN ESTIMATE

A variety of forms are useful in preparing an estimate. Some of the more common forms are:

- take-off sheets
- take-off summaries
- estimate worksheets
- estimate summary

Take-off sheets and summaries vary from company to company, whereas estimate worksheets and summaries are usually standard. Some of the most common take-off sheets are designed for voluminous, repetitive work such as piping, electrical, and civil work. As a minimum, the following sheets should be designed:

- piping quantities
- fittings lists
- valve lists
- piping supports
- electrical raceways
- rebar lists

An estimate worksheet that is common in industry is shown in Figure 8.1.

The use of computers in preparation of estimates has had mixed success, primarily because the time saved in their use cannot offset the cost for their purchase and maintenance. While it can make the estimator's job easier, not much saving in his manhours can be expected.

ESTIMATE REVIEWS

Prior to acceptance of an estimate, it should be reviewed. This is best done in separate review meetings with engineering, construction, and management. In each case, the estimate must be reviewed line by line to ensure that the scope, as envisaged by the estimator, is agreed upon and that the costs estimated are acceptable.

Then a combined meeting is necessary. At this meeting, the estimate must be presented, after having been adjusted as a result of input from the previous meetings. Once the estimate is agreed upon, a budget must be struck that will become the basis for cost control. Budgets to be established are:

- equipment prices
- rates and manhours for labor

FIG. 8-1 — ESTIMATE WORKSHEET

ITEM AND DESCRIPTION	QUAN-TITY	UNIT	UNIT COST		MANHOURS			TOTAL COST			
			MAT'L	S/C	UNIT	TOTAL	$/MM	MATERIAL	LABOR	SUB-CONT.	TOTAL

- quantities allowances for materials
- subcontract packages and rates
- engineering rates and manhours
- construction management rates and manhours
- rates for construction equipment and tools
- costs for miscellaneous services

Once these budgets have been developed and agreed upon, the project can continue and its costs and schedule be controlled.

CHAPTER

9

COST CONTROL

Okay, you've got a commitment from the board for a multimillion dollar expense to engineer and build a plant. Now what? How are you going to be sure that you can control the spending day by day?

CODES OF ACCOUNTS

The first thing you had better do is see that the code of accounts framework that was set up by your estimator adequately covers your needs. Second, make sure that a method of assigning codes to expenditures and commitments is set up. One word of advice gleaned from bitter experience: *Do not depend on engineers, construction superintendents, purchasing agents, inspectors, or anyone else to code expenditures or commitments.* Two and only two groups or persons should be responsible — estimators and cost controllers. Otherwise, it will be done inaccurately and an enormous amount of time will be spent correcting computer printouts, purchase documents, contracts, you name it.

A code of accounts breaks down the project into cost centers. This is done by area or by function, and a flowsheet or plot plan is used to draw up the basic breakdown. Stem accounts are assigned to these basic areas. If we use a steel mill as an example, the stem accounts could be set out as follows:

    100.0   Service facilities and infrastructure
           110.0   Site preparation
           120.0   Roads
           130.0   Railroads
           140.0   Service buildings
           150.0   Water supply
           160.0   Sewage disposal
           170.0   Electrical supply and distribution
           180.0   Service buildings
    200.0   Scrap yard
           210.0   Sorting yard and equipment
           220.0   Cutting and shredding
           230.0   Plant feed system
    300.0   Melting
           310.0   Charging system
           320.0   Electric arc furnace
           330.0   Ladle handling
    400.0   Continuous casting
           410.0   Tundish area
           420.0   Straightening and cutting
    500.0   Environmental control
           510.0   Baghouse
           520.0   Fans and ducting
           530.0   Stack
           540.0   Cooling water
    600.0   Indirect costs
    700.0   Project management and engineering

Once the stem accounts have been set up, then the work is further broken down into decimal accounts to the degree necessary to identify and control costs. One should keep in mind that the KISS (Keep It Simple, Stupid) principle applies: The simpler the better, commensurate with quickness and ease of retrieving costs. Don't forget that cost control is a control function, not an accounting function, and that fine detail may not always be necessary or practicable. Cost reporting and cost control are not the same thing!

Decimal accounts typically break down into: earthwork, foundations, structures, siding, roofing, mechanical equipment, piping, electrical, instrumentation, and so on and are then further broken down as needed. If, for example, foundations are under decimal account 0.2, then that account may be further broken down to:

.22	Structural foundations
.221	Forms
.222	Rebar
.223	Concrete
.224	Anchor bolts

COMPUTERIZATION

If ever a success has been made in project management, cost control is it! Whether your project is small or large, insist on access to a computer. Normally, a desk-top computer, such as the Apple IIe, IBM Peanut, or Commodore VIC 20, is adequate for control of most projects. However, access to a mainframe computer may be desirable if you are handling a very large project or many projects at the same time.

Cost control involves repeatable calculations that change with input on a routine basis. While manual calculations have been done for years, the process has been fraught with frustration when things just don't add up due to errors and omissions.

In cost control reporting, the printing and distribution of hard copy tends to become unwieldy due to the limitations of present day printers. Reams of reports get distributed and, as a result, are sometimes not read. Exception reports are better, or, at the very least, reductions can be made on a photocopy machine to facilitate handling by the recipients.

WHAT ARE WE CONTROLLING?

If you look back at Chapter 8, the main bulk of capital costs is in direct costs. This is further subdivided into materials, labor, and subcontracts. Materials can be further subdivided into bulk materials required for construction and equipment. Let's take each of these direct costs, one at a time, and decide how to monitor and control them.

Starting with materials costs, the obvious place to begin is with equipment. Most equipment is bought with individual purchase orders. Although spending for each piece of equipment may be over a period of time, with progress payments and the like, the essential action that takes place when a purchase order is let is that a commitment has been made. The most important step, then, in reporting costs is the equipment commitment report (Figure 9.1), which can be organized by cost code or purchase order number or both, depending on the computer capability. The cost controller receives his information on copies of the purchase orders, the cost coding of which has been applied by the cost controller. Since the longer the project goes on, the more inactive the listings become, it is best to print out an exception report, perhaps just showing "this period" activities and activities that are still "to go." Otherwise, the recipients of the report, especially if there are a large number of purchase orders and a wide distribution of the report, will receive large quantities of useless paper.

Bulk materials costs are handled the same way, i.e., by a commitment report, except that organization can only be by cost code. Control of bulk materials, such as piping or electrical cable, usually includes quantities, unit prices, and total costs. Information is reported by the purchasing department, either in the engineering office or at the site in the case of field purchase.

Labor, too, must be reported by code number. Since labor has been estimated by activity, then it follows that time sheets should reflect the proper codes. These are usually coded by a cost controller in the field, in accordance with foremen's reports.

Subcontracts are reported in a procedure similar to that for equipment purchases except that information is provided by the subcontractors. Each subcontract is reported separately in a contract status report (Figure 9.2), and its results are in turn reported on a subcontract summary, which is organized by cost code.

Now, having provided the means for reporting costs and *getting them reported quickly*, the main function of the cost controller is to *predict where costs are going*. Knowing where they are going

FIG. 9.1 — EQUIPMENT COMMITMENT REPORT
SINKING SHIP STEEL MILL

DATE: **FEB. 29/88**
AREA: **SCRAP YARD**

P.O. NUMBER	DESCRIPTION	COST CODE	BUDGET
4123.001	CONVEYOR HARDWARE	220.5531	75,000
4123.002	ELECTRIC MOTORS	VARIES	125,000
4123.003	O/HEAD CRANE	210.4710	19,000
etc.	etc.	etc.	etc.
		TOTALS	1,279,000

COMMITTED THIS PERIOD	TO DATE	VARIANCE
8,000	68,000	(7,000)
15,000	156,000	31,000
18,500	18,500	(500)
etc.	etc.	etc.
135,500	1,043,000	(236,000)

or likely to go is largely dependent on trend forecasting, a subject which is treated in detail in the next chapter. *A trend forecaster monitors potential changes in scope, whereas the cost controller is responsible for analyzing trends in predicted costs.* For example, if productivity in the installation of brick walls is not as high as originally estimated, then this fact should be illuminated by the cost controller. Similarly, progress in engineering or other facets of the project must be examined on a regular basis and reports made to the project manager.

A cost controller must also look at the project on an overall basis. When the project is first conceived and its costs are agreed upon, a time frame for spending the funds is established. Once all the studies necessary to finance the project are completed and a decision is made to actually spend money on hard costs, then, if it is done properly, the cost, schedule, and scope of work are frozen and the process begins. Although money is being directed towards a known target and the project is progressing towards a

FIG. 9.2 — SINKING SHIP STEEL MILL

Record of Contract / Status

Job: _____ No. _____ Code No. _____

Sub-Contractor: _____ Trade: _____

_____ _____

_____ _____

K.P. : _____ tel: _____

holdback: _____

Contract Amount:		**Progress Draws:**			
		No.	Date	Amount	Cum. $
Original:	$ _____				
c/o# _____	$ _____				
c/o# _____	$ _____				
c/o# _____	$ _____				
c/o# _____	$ _____				
c/o# _____	$ _____				
c/o# _____	$ _____				
c/o# _____	$ _____				

known completion date with, perhaps, a few milestones along the way, the rate of spending money is often ignored, and whether the project has succeeded in keeping to the budget is not known until the final count is in. Computer printouts showing budgets, commitments, and spending are worthless if a majority of the expenditures take place at the eleventh hour. It is necessary, therefore, to establish milestones for costs as well as schedule and to monitor progress accordingly.

S curves for spending and commitments should be drawn up for plotting the progress of the project on a monthly basis (Figure 9.3). The S curve makes the assumption that it takes a while to get going, therefore the slope of the curve increases gradually. Similarly, completion of a project reverses that trend and the rate

FIG. 9-3 — S CURVE OF EXPENDITURES

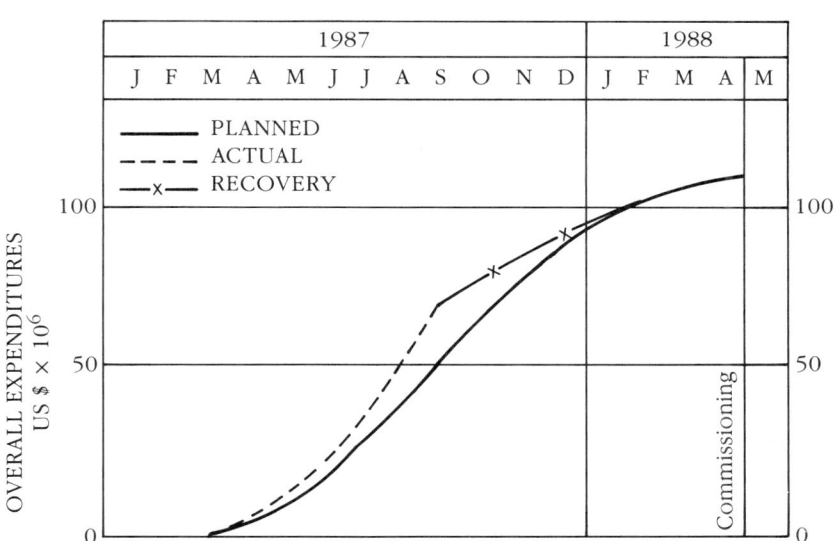

of spending tapers off at the end of the project. In between, the curve has a more or less constant slope. This is typical of capital spending and can serve as a rough guide. The actual progress is plotted on a periodical basis and is compared with the predicted curve.

The reason for going to an S-curve rather than calculating an actual spending path prediction is that it simply cannot be done more accurately without creating a bureaucratic monster that is as difficult to control as the project itself. It is somewhat ludicrous, however, to expect a project to follow an S curve exactly, as that puts the project manager — unless he is very lucky — in the position of always being off the curve. In the interest of reality, it would be wiser to draw an S-curve envelope (Figure 9.4). Also, the S-curve configuration can be amplified by preparing a milestone spending list to coincide with a milestone schedule (Figures 9.5 and 9.6).

FIG. 9-4 — S-CURVE ENVELOPE OF EXPENDITURES

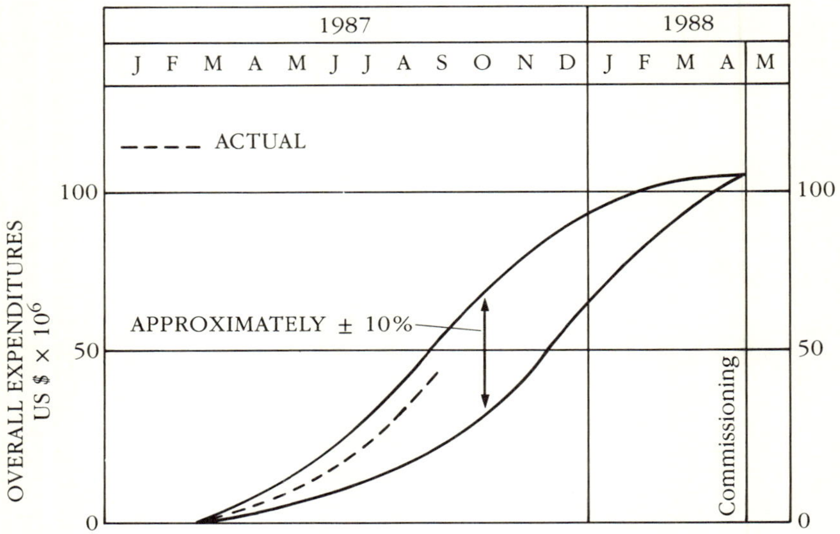

FIG. 9-5 — MILESTONE SCHEDULE

Bald Mountain Billiard Ball Beneficiation Plant

Milestone Event	Date
Kick-off meeting	March 1, 1987
Process design frozen	May 1, 1987
Place order for mills	May 15, 1987
General arrangement approved	June 1, 1987
Order transformers	June 15, 1987
Award contract for foundations	July 1, 1987
Award contract for building steel	July 15, 1987
Enclose building	October 15, 1987
Install elutriators	December 15, 1987
Complete piping installation	February 15, 1988
Complete wiring	March 31, 1988
Complete instrumentation package	April 15, 1988
Start commissioning of equipment	May 1, 1988
Dry test and equipment acceptance	June 1, 1988
Load testing	June 15, 1988
Start up	July 1, 1988

FIG. 9.6 — MILESTONE COMMITMENT SCHEDULE*

Bald Mountain Billiard Ball Beneficiation Plant

Event	Cost	Date
	$ x10^6	
Award engineering contract	$ 8	March 1, 1987
Purchase mills	$ 14	May 15, 1987
Purchase transformers	$ 1.5	June 15, 1987
Award contract for foundation	$ 3	July 1, 1987
Award contract for building steel	$ 20	July 15, 1987
Purchase polishing equipment	$ 5	July 31, 1987
Award mechanical contract	$ 12	October 1, 1987
Purchase control panels	$ 1	February 15, 1988
Award electrical contract	$ 0.5	February 15, 1988
TOTAL	$ 75.0	

*Note that the above milestones cover about two-thirds of the total amount that will be committed and serve as a guide for spending money.

CHAPTER

10

TREND FORECASTING

A project manager for an engineering firm who has been running a mill expansion job for a mining company has just had a change order put in front of him for approval. It is for the addition of a second rotary dryer in the flotation circuit. It will cost $1.3 million, which is exactly $750,000 over his budget. "I can't approve this," he tells his design manager.

"What?" says the design manager indignantly. "It's *got* to be approved!"

Restraining an urge to remind the design manager just who is boss, the project manager says quietly, "What do you mean?"

"All the drawings have been done for the additional dryer. If we have to change back to the original design, it will lose us three months on our schedule. That will cost the client over a million dollars in lost production. What's more, it will blow our engineering budget. It's already been stretched by adding the dryer."

The project manager blanches. He knew about the additional dryer, but he had no idea that it would cost so much, especially now that he's running so close to budget. "How did we wind up with an extra dryer?" he asks.

"Client's chief of maintenance asked for it," the design manager replies. "He popped in here one afternoon a few months back. Said we'd be fools if we didn't add another dryer. So we put one in. I told you about it, remember?"

The project manager nods. He remembers, all right, but he never thought it could have such dire consequences. "Thank goodness the client asked for it," he thinks. "That puts us off the hook."

"Wha-at?!!" says the client's project manager, fuming. "I never authorized an additional dryer. What's more, we tried that already in one of our western plants. Made it easy as hell for our maintenance people, but it didn't improve the bottom line. Made it worse, in fact, much to our embarrassment. You had better do something about it."

Explaining the time-and-cost restraint they are in is bad enough, but it also creates a shock wave throughout the client's organization.

"I never *ordered* them to put in the dryer," says the chief of maintenance. "I merely suggested it, expecting that it would go through the regular approval channels. Either it would get approved — in which case my future job would be easier — or it would get chopped."

He fails to mention that he had hoped that it would slip through.

The consultant now feels as though he has been sold down the river by the client. The client feels that the consultant is just grabbing an opportunity to get extra manhours. Bad feelings begin to evolve, but nothing like what develops when a request for extra money gets to the board.

"Wha-at?!!" exclaims the chairman. "We've already allocated all of our resources based on your definitive estimate. The design was supposed to have been frozen. We're stretching our credit line at the bank already. You had better *do* something, because the extra money is *not* approved!"

Now the client's project manager is furious with his chief of maintenance (who is secretly pleased) and even more so with the consultant. Yet, he feels responsible. As a manager, he should have kept his finger on the pulse of the job, known about the dryer. Obviously, his line of communication within his own organization is no better than the consultant's.

Exactly the same thinking process now occupies the mind of the consultant's project manager. Negotiating a major installation contract has preoccupied him during the past few months, but he knows that this is not a good enough excuse. Like the client's project manager, he knows that his job is on the line.

Emergency meetings into the wee hours make it clear that the momentum built up is too great to eliminate the extra dryer from the design. It will have to stay. The $750,000 overrun in the budget will have to be eliminated elsewhere in the project. That is possible by trimming scope in other areas and will result in a plant that is a Cadillac in the drying section but only a Volkswagen in others. If other cost upsets come in, some sections could be reduced to go-carts, truly an unsatisfactory design. It will be the last engineering project assigned to that consultant by the mining company. Promotions within the projects group are likely to get pretty thin, too, so the client's project manager starts looking around for another job.

How would trend forecasting have prevented this situation from developing?

TREND FORECASTING DEFINED

First, let us define trend forecasting in simple terms. Trend forecasting is an early warning system designed to advise top management of possible changes in cost and/or schedule with sufficient advance notice that the changes either can be easily handled or corrective action can be taken. While trend forecasting is a simple process, it is invariably misunderstood and often ignored or misused. *Yet it is the single most effective communications tool in the industry!* One has to ask the question, "Why is it misunderstood?"

Perhaps it would be better to explain what trend forecasting is *not*. The name itself is misleading because project people tend to think of trends as changes in cost or schedule that are evolving as a result of unknowns. They simply happen: things are costing more (or less) or are taking longer (or less time) than predicted. It is thought that by plotting the progress on a curve, like a construction S curve, the trend can be predicted or forecast. The difference is this: An S curve tells you where you have been; a trend forecast tells you where you will go unless you quickly decide to change your course. The former could be referred to as trending, the latter as trend forecasting.

It would seem logical, because of the confusion sometimes resulting from the name, to come up with a better designation for this process. Suffice it to say that it has been tried. The present name has been debated, cursed in great detail, and left to stand as it is.

THE TREND FORECASTER AS DEVIL'S ADVOCATE

Who should be a trend forecaster? If we place importance on the system of trend forecasting in a project, then it becomes imperative that we assign the right person to the job. The trend forecaster should be at least as strong as the project manager as regards knowledge of the project and the project management process. He should be the project busybody, the snoop, the person who knows everybody's business — the project manager's, the client's, the engineers', the estimators', the schedulers'. He should be a strong estimator and, ideally, should have grown from the estimating group. He has to be able to come up with order-of-magnitude costs — quickly — and if he cannot, then he should be able to badger someone else into providing them. He is a tattletale. He has to be able to squeal on his own project manager if necessary in order to produce a trend forecast. He has to be able to sniff out trends at meetings, conferences, from memos and correspondence, and on the floor. He should be nearly the last, if not *the* last person to speak at project meetings, asking the question, "Are there any more trends?" In our example at the beginning of this chapter, he should have identified the extra dryer as a trend, put a cost and schedule to it, and raised it in his weekly trend forecast for all to hear — from the engineers through to the client. It would have stimulated an earlier decision.

WHEN AND HOW TO TREND FORECAST

Trend forecasts should be made weekly within the project and should be presented at least monthly to the owner and quarterly to the board. Frequently, an astute project manager will raise a trend forecast as a trend becomes apparent if he has a suspicion that it will create waves in the client's organization, or, quite simply, if he is anxious to get a reading on the reaction of the powers that be to an idea.

Figure 10.1 shows a typical trend forecast, using our earlier

FIG. 10.1 — GRANDIOSE ENGINEERING COMPANY TREND FORECAST

WEEK NO 47
DATE: July 11, 1987
CLIENT: Bottomless Mining Company **PAGE** 1 OF 1
PROJECT: Pisswilly Falls Concentrator

TREND NO.	DESCRIPTION	ENGINEERING MANHOURS	SCHEDULE	CAPITAL $ × 1000
47-1	Add rotary dryer to flotation section, required to eliminate the effects of downtime for maintenance.	+500	No change	+1,500
	TOTAL TRENDS	+500	No change	+1,500
	BUDGET	150,000	36 months	225,000
	LATEST FORECAST	155,000	36 months	224,250
	OUTSTANDING TRENDS	none	none	none
	NET EFFECT OF TREND	+500	No change	+ 750
	TREND FORECAST	155,500	36 months	225,750

THIS FORM SHOULD BE PRINTED ON COLORED PAPER FOR CONTRAST

example. A trend can be positive or negative and can indicate manhours, costs, and/or schedule. Even a zero trend can be raised if a proposed change in scope does not involve cost or schedule changes but should be widely known through senior management.

Rotary Dryer Example Revisited

Using our previous example, let us replay the scenario as though trend forecasting had been in place since project inception.

Our trend forecaster, a sneaky little devil, spots the client's chief of maintenance in the office reviewing drawings with the design manager. He decides to eavesdrop on the informal meeting — a mandate that he has, by the way — and he overhears the chief's request for an additional dryer. "How much does an extra dryer cost?" he asks.

"Oh, the others were about a million each," says the design manager. "Maybe a little more with the instrumentation. Add another half million for the extra building space and foundations."

"What about schedule?"

"No problem if we start right in on it and do a little overtime in the engineering department. I'll let the project manager know, although he'll be hard to get hold of. He's negotiating contracts right now."

"Manhours?"

"About five hundred more. Peanuts, really, but it could be touchy. I'm over budget now."

"Why don't I raise a trend?"

"Good idea. That way we can gamble — start right in on the work. If it gets turned down, we won't have wasted a lot of time on it. It's going to take us a month before we have enough information to prepare a change order, anyway."

They agree. The project manager sees right away that it would put the project over budget, but if that's what the client wants, it could be good for the project. When the trend forecast is presented, however, the client's project manager squelches the idea, based on the company's experience at their western plant. If he had decided to go with the extra dryer, he would have immediately passed on the trend forecast to the company president, who would decide whether or not it was worth asking the board for more money. If he decided it was, then the board would be advised that a change order could be raised in the near future. At that point, the change order still could be declined, based on cost or new information, but at least the management chain would have been forewarned.

In general, a trend forecast need not be approved by anyone, only acknowledged, which in effect is tacit approval to proceed to the change order stage. That is precisely what gives the trend forecaster license to broadcast trends on the skimpiest of information. There should not be the slightest excuse not to raise a trend, since by definition a trend is *not* a change, only the possibility or probability of one. In our case, further engineering could have revealed a better solution to the extra dryer, resulting in a negative trend or an outright cancellation of the earlier trend.

WHEN SHOULD TREND FORECASTING BEGIN AND END?

Obviously, trend forecasting is of greatest value in the early stages of a project, when the scope is looser than, say, after the definitive estimate. You can begin trend forecasting as early as when the idea of a project is conceived. In fact, when a project idea is first bandied about the office, a value invariably gets put on it through long- or short-term forecasts or whatever. This value tends to stick in people's minds, sometimes becoming stubbornly fixed. Personnel are often shocked when, sometimes years later, changes in scope have taken place. By trend forecasting on a regular basis, these shocks can be eliminated.

When should trend forecasts end? It is popular to stop at the definitive estimate stage, because the scope is supposedly frozen at that point. It would be prudent to continue the process for a few months longer, however, to ensure that the decision makers were serious in freezing the scope of the job.

CHAPTER

11

PROCUREMENT

Quite often the terms "purchasing" and "procurement" are used interchangeably. In project management, the purchasing function ends with the placing of a purchase order. The procurement process covers the whole gamut, from soliciting of bids to delivery and warehousing of goods at site. Purchasing is a small part of procurement.

THE PROCUREMENT CHAIN

The procurement chain involves soliciting bids, placing purchase orders, inspecting, expediting, transporting, accepting, and storing and handling at site. It also includes final acceptance at start up as well as claims on warranty.

Purchasing

Most projects have the majority of buying done in the engineering office, with field purchases kept to a minimum. In either case, the process is essentially the same. Except on very small projects, a buyer or purchasing agent is assigned this task.

Invariably, there are moments of tension between purchasing agents and engineering, no matter what their relative positions in the hierarchy or matrix. Engineers consider purchasing agents to be servants of the engineering team and vice versa — the purchasing people prefer to use the engineers to provide specifications and data so that they can perform their duties efficiently. In practice, these tensions usually resolve themselves and the job gets done.

The purchasing function usually begins with publication of an equipment list by engineering, from which a bidders list is compiled. This list consists of favorites of engineers, clients, purchasing agents, and friends of brothers-in-law's girl friends. Prequalification perusal should be performed as a team, otherwise the number of bidders per proposed purchase order can grow to endless lengths. Except in some rare cases, it does not make sense to solicit bids from more than four or five suppliers for a given item. It is not fair to have bidders submit a lengthy bid package if their chances of getting the order are slim. In any case, the master bid list should be published as a guide only. Additions and deletions can be made as the need arises.

An important function of the purchasing group is to ensure that bidders on the list are in healthy economic shape. One mining company purchased an on-stream analyzer at a bargain, only to find that when it would not work, the manufacturer did not have the financial resources to fix it. A paper mill was nearing completion in Texas when delivery of the paper-making machine could not be made, owing to the bankruptcy of the manufacturer. At the bid list preparation stage, it is normal to include suppliers who are known by reputation and through previous dealings. Then a thorough commercial check is made prior to recommendation for purchase. A cursory check prior to soliciting bids can save some time, however, if suspicions arise.

Suppliers who have successfully made the master bid list usually have helped the engineers and estimators in preliminary design at no cost. It is normal for the project team to return the courtesy.

General conditions are normally a standard company attachment to purchase documents, but sometimes they are custom designed for a particular project. They are meant to cover the legalese and gobbledegook that are common to all purchase orders, and they may be prepared with the help of a lawyer. They used to appear in fine print on the back of purchase orders;

however, of late it has become necessary to put them front and center so that, in the event of litigation, you cannot be accused of hiding them. Items covered include statements on acceptance, delivery, insurance, title, taxes, confidential information, warranties, conformance with laws and codes, patents, overdue accounts, default, modifications, waivers, termination, and liability. Suppliers usually have their own general conditions with their bids, although the similarities between the two sets of conditions often outweigh the differences.

Requests for bids begin in engineering, where specifications are written. A requisition that summarizes the scope of work in the specifications is prepared. The codes of accounts are added by the cost controller at that point. Then the package is sent to purchasing in a memorandum. Purchasing checks it through to see that it's complete and sends it out for bids with a covering letter. On government projects, bids must be received at a specified date and must be sealed. They usually are opened publicly to ensure that no graft or collusion has taken place.

In commercial or industrial projects, public openings are rarely held. Instead the purchasing agent is usually anxious to open the first bids in, to see if the suppliers have interpreted the documentation correctly and submitted complete bids. At that point, a preliminary bid analysis and commercial check on the low bidder are made, and all of the bids are sent to engineering.

During the bid analysis stage, the purchasing people and the engineers are likely to work very closely together. It is rare if all bids are correct, and a fair amount of contact between suppliers and engineering is required. It is good practice to resist isolating the engineers and suppliers from purchasing and vice versa at this stage so that everyone tells the same story, thus minimizing misleading statements.

A typical bid analysis is done in two sections: commercial and technical. The commercial section compares quantities, delivery times, FOB points, basic prices, freight costs, insurance costs, drawing costs, warranties or guarantees, and costs for spares. The technical section compares the technical quality and checks for conformance to specifications. Usually the lowest bidder who conforms to specifications is recommended, unless there are other factors, such as lack of compatibility with other equipment, guarantees, delivery time, quality of work, and so on, that work against him.

More often than not, clarification of minor details is required prior to making a commitment for purchase. Also, it is desirable to have a meeting with the successful bidder prior to giving him an order, in case any second thoughts develop. And sometimes a trip to the manufacturer's plant is helpful in reinforcing the decision to place the order with them.

One pulp and paper consulting company is well known for having a final meeting with the three lowest bidders. Their technique is to have a supplier in each of the three waiting rooms and to bring them in one at a time to see how much they will reduce their prices. Naturally, most reliable companies know that this type of bargaining will eventually happen and build an extra amount into their prices so that they will be comfortable in making a cut. It's a game of nerves, but only a game, one that nobody really wins.

Approval for purchase should come from the project manager. Preparation of a purchase order should begin with engineering, who will write a request for purchase. If this is done properly, the request for purchase will become the basis of the purchase order, for it will contain the same scope of work that originally appeared in the request for bids, except that specific reference to the suppliers bid and any approved exception to the original specifications will be made. It is not a bad idea to revise the specifications to conform to the purchase order and quotation at this stage, if time is available to do it.

It is usual to buy spares for the first year of operation at the same time as the order is placed for the equipment. This should be done on a separate purchase order, since spares are not capitalized as far as taxes are concerned.

It would be foolhardy if follow up was not done after placement of an order. Salesmen make promises that are not necessarily passed on to the floor of the manufacturer. Two basic things can go wrong after placement of an order: 1) The equipment does not conform to specifications or quality expectations; and 2) It will not be fabricated or manufactured on time. That is why inspectors and expeditors are needed.

Inspection

It is quite important that you plan ahead for the inspection function. First of all, it should not be necessary to inspect each and every piece of equipment. In a process plant that uses a lot of small pumps, for example, there may be a case for waiving the inspection, because a lot of spare pumps will be available at site should one of them not be up to scratch. It could be sent back without a lot of panic ensuing. But it is also a question of planning the schedule of the inspection department so that inspection is not rushed. It is always a good idea to specify which equipment needs to be inspected, and in some special cases to what degree, at the time of going out for bids. If this is indicated on the body of the request for bids, then both the inspection department and the bidders are forewarned.

Although every effort should be made to deal with reputable suppliers, you should never assume that the equipment is going to be produced and delivered to your satisfaction. Manufacturers are run by people, and people are not infallible. One major US company built 52 drum-type magnetic separators for a mining company. The inspector followed their manufacturing process, inspecting it from time to time and accepting it at various stages. The final assembled separators were rejected by him, however, causing quite a flap, both at the site, where late delivery would hurt the critical path, and in the factory. A foreman, it seems, had decided to hammer the rolled stainless steel shells into shape prior to welding them closed. That left the drums wavy at the seams, reducing their effectiveness as magnetic separators. Replacing the shells was out of the question for the manufacturer due to the high cost of stainless steel. He elected to remove them and correct them by hand — also an expensive process. But the problem got resolved.

Very specialized inspection and tests that must be witnessed should also be flagged at the time of bidding. Pressure vessels should be carefully tested, even though they are expected to arrive on site bearing official stamps of approval. Incidentally, it is important to take rubbings of the code stamps and to keep them on file as proof of having had the vessels inspected and properly documented. Performance tests are sometimes required on specialized machinery, since it would do precious little good to have it arrive on site, be installed, and then not produce. In no case should the manufacturer be allowed to ship without written approval by the inspectors.

Dimensional checks are an important inspection function, but they obviously cannot be carried out in all cases. Naturally, equipment that does not fit in the field can be a major problem. An attempt should be made to identify key dimensions. For example, bridge girders could probably be modified in the field if they were delivered with the wrong dimensions, but the costs can be enormous if rented cranes and crews are standing by while the work is being done.

Inspectors have to develop a good working relationship with suppliers and manufacturers. An atmosphere of trust will make it easier to gain access to equipment and manufacturing facilities. It is not unknown for an inspector to nitpick, nor is it unheard of for a shop foreman to cover up deficiencies. Switching name-plates is an old trick that fools an inspector into believing that the equipment inspected is the same as that which will be delivered to site. A good working relationship will minimize such skullduggery.

Expediting

"Will you get that sonovabitch expeditor of yours off my back?!" cried a manufacturer of slurry pumps over the telephone.

"Why?" asked the project manager.

"He's trying to get me to start manufacturing your pumps."

"So? That's his job. What's the problem?"

"I haven't even received the purchase order yet!"

One characteristic of a good expeditor is zeal — his job is to see that goods are manufactured and delivered on time. Obviously, in the example above, the expeditor had received a copy of the purchase order before the manufacturer.

Suppliers often make delivery promises that become difficult to keep. Sometimes it is simply a case of a salesman making a commitment without consulting the shop foreman. At other times the manufacturer may receive two or three orders at once, much to his delight but to the chagrin of the foremen who have to try to live with the commitments that have been made. In this latter case, the expeditor who makes the most noise is the one who is most likely to have the goods delivered on time. It is very common to have one job put aside to the advantage of another.

Expediting does not just involve the manufacturing process. In the beginning, vendor drawings for approval must be chased both at the supplier and at the engineering group where they are being processed. And it does little good to have equipment manufactured on time if trucks or railroad cars are not available for shipment. The shipping cycle must be followed from loading to enroute monitoring to offloading and storage. Your expeditor may wave goodbye to a departing train at the factory, but his smile will fade if the boxcar of goods gets shunted onto a siding in the Chicago railyards for a week. And if it finally arrives at site but nobody has arranged for an offloading crane or a storage area, then demurrage costs will mount. Careful planning, of course, is the key to all of this, but it can be to no avail if an expeditor is not there to follow through.

Vendor Print Control

If a machine arrives at site and it is two feet longer than the space allowed, or if nobody realized that it needs a substructure, then corrective measures can be very costly. Engineering depends on drawings of the equipment in order to design and build foundations or structures, and the inspectors depend on drawings for checking. Except in rare cases, therefore, manufacturing or delivery should not be allowed to proceed until engineering has approved drawings that are certified as correct by the supplier. Even off-the-shelf equipment can have design or dimensional changes made of which engineering should be aware.

Problems occur in an engineering office when an orderly and expeditious handling of vendor drawings is not set up. Drawings come in and wind up sitting on someone's desk for days on end. Meantime, the manufacturer cannot proceed. To overcome this, set up a vendor print logging and monitoring system. Figure 11.1 is a typical log sheet. There should be one for each purchase order or piece of equipment.

When drawings are logged in, a stamp (Figure 11.2) is applied and the drawings are routed for approval. Some time can be saved by distributing copies to each group rather than circulating one set sequentially. Many companies put a disclaimer on this stamp, for example, "Manufacturing may proceed. This approval does not relieve the manufacturer of the responsibility for the accuracy . . . . " In other words, if anything goes wrong, even if the engineer has asked the supplier to do something wrong on the drawings, then the supplier is to blame. Oh well, why not?

Installation and Commissioning

Part of the procurement chain includes the use of experts from the manufacturer to assist in installation and commissioning of his equipment. Normally, a few days have been allowed for this by the manufacturer. After all, it is in his interest to see that these

FIG. 11.1 — VENDOR PRINT CONTROL LOG

Project No. 1234
Project: Snake Eyes Dice Factory
P.O. No. 1234M-07

Description	Drawing No.			1
Dice Loader Gen. Arrt.	1234-M-07-02-01	IN	A	15/12
		OUT	2	27/12
Lead Melter Mechanism	-02	IN	I	15/12
		OUT	1	27/12
Foundation Layout	-03	IN	A	29/1
		OUT	1	10/2
Wiring Diagram	-04	IN	A	29/1
		OUT	1	10/2

Date: Feb. 10/99
Equipment No.: 1234M-07-02
Description: Dice Loading Machine

STATUS / REVISION NO.								REMARKS
	2		**3**		**4**		**5**	
C	29/1							
1	10/2							

STATUS CODE:

A = For Approval
C = Certified for Construction
I = For Information, Approval not Required
N = Not Approved
1 = Approved for Manufacturer
2 = Approved as Noted — Correct and Resubmit

FIG. 11-2 — DISTRIBUTION AND APPROVAL STAMP

Dept.	Initial	In/Out		Approval Status:
Mech.				0 ☐ Not Approved
Civil				1 ☐ Approved
Elect.				2 ☐ Approved as noted. Resubmit
Inst.				N ☐ Approval not Required

Approved by: _____

Date: _____

Vendor Print No. _____

things are done properly. However, it is usual to have a per diem rate for extra work included in the original bid package.

Acceptance of equipment in the last stages covers four phases:

- as delivered
- as erected
- as commissioned
- as rated

Naturally, if equipment arrives in a state of disrepair or with parts missing, there are grounds for dispute. It is fair to provide the supplier with notification that his equipment has been delivered as ordered.

It behooves the construction manager to seek the manufacturer's blessing on some installations, particularly where interconnecting machinery, ducts, chutes, or piping could affect the performance of his equipment. Commissioning is essentially a checking-out stage. Often the manufacturer is required to tune up, calibrate, or adjust his equipment prior to start up. He should provide a certificate signifying that this has been done to his satisfaction. And finally, if the equipment starts up and operates satisfactorily, a notice of that fact should be given to the manufacturer.

The above, of course, applies to major process equipment. For the myriad of small pumps and valves and fans, the supplier and user are normally satisfied by silence from each other.

CHAPTER

12

SPECIFICATIONS AND CONTRACTS

Specifications and contracts do not get written. They get perpetuated from one project to the next. Everyone who comes in contact with them has to add something clever, sometimes the result of burned fingers, and over the years a simple covenant between two or three parties can become an epistle of many volumes. The popular notion of these documents is that they are designed to protect the issuer from whatever goes wrong, no matter what.

Something seems to drive specification writers to produce documents that are almost unintelligible. Gobbledegook prevails where clear understanding is needed, and the more legalese, the better. A lot of specification and contract writing is a cut-and-paste exercise, and nobody can be blamed for taking advantage of previously tried and proven documents. An engineer once borrowed a paragraph from a contract that had been written by a lawyer. It was a completely redundant section on "enurement," written in rhetorical legalese. Admittedly, it was added for bulk — the contract appeared a bit thin without it. Naturally, the definition of "enurement" needed some research. However, when it came time to sit down and negotiate the contract, the engineer felt thoroughly embarrassed when he realized that he had forgotten what the word meant! Fixing an unflinching gaze on his counterpart, he implied that *anyone* ready to sign a contract should have researched his facts, and for goodness sakes, let's get on with it!

Words, like *enurement*, with which few people are familiar, are known as weasel words. Anything designed to help you wiggle out of responsibility falls under the same category. And fine print is a definite no-no.

Specifications and contracts serve two basic functions: 1) to set out the respective agreements and responsibilities of the parties involved; and 2) to protect the parties from each other in the event that things beyond their control go wrong. The intent of a contract is to satisfy all parties to it, not to satisfy one party at the expense of the other. Too often people approach a contract wearing cast iron trousers, as if the protection afforded them by the contract were more important than the pledges the parties have made to each other. When that begins to happen, one should ask whether or not the contract should be made at all, for *the essence of any contract is the trust behind it*. If you don't know who you are dealing with, find out. Dealing with reliable people is better protection than all the paper in the world.

SCOPE OF WORK

How often do we read a specification or contract and have to dig through hundreds of pages and thousands of words just to find out what the scope of work is? What, after all, is more important than telling the contractor or supplier what he is expected to do and what he is not expected to do? Then, to add insult to injury, we find a statement that reads "The following is included, but not necessarily limited, to . . . ."!!

In writing specifications and contracts, if you put the scope of work up front and center, and if you are very simple, clear, and concise, you will have come a long way towards effecting a successful agreement.

STANDARDS AND STANDARDIZATION

Codes and standards are a very necessary requirement in some cases, particularly where hazards are involved. If you wish to show your ignorance, however, you will ask for conformance to a list of codes or standards that are as long as your arm. Normally, only a couple of codes, such as ASME and API, are necessary to assert the intent of the specifications. Any others that the supplier may conform to can be stated in his proposal at your request.

Standardization, by contrast, is an attempt to portray standard approaches to all specifications or contracts on a specified project or even throughout the company. If, for example, all electrical devices and motors are to be explosion proof, then it makes sense to state that on a standard sheet on electrics attached to all specifications and contracts that have electrical devices, be it a mechanical, civil, or architectural package.

The trouble with that approach is that standard packages have a tendency to grow in size. If, for example, you have a belt conveyor contract, then it is conceivable that you could add an electrical standard, an instrumentation standard, dust collection standards, coupling standards, guard standards, V-belt standards, gearbox standards, motor standards, and platework standards, all with introductory blurbs, gobbledegook, and disclaimers, ad nauseum. Invariably the contract package becomes several inches thick. When it is issued, the person in charge is not likely to check standards in detail, because they're standard, right? And the chances for error escalate. When all of this stuff gets added to the procurement package (request for bids, general conditions, vendor print requirements) it becomes a very hefty document indeed. When it finally gets issued, the purchase order often carries the same set of documentation as attachments.

On one project, a supplier received a half-inch-thick purchase order, followed smartly by a telephone call from an expeditor enquiring as to why the wiring diagram had not been forwarded. The purchase order was for an anvil! Someone added a standard to the specifications that didn't belong, and you know that when an operation becomes that bureaucratic, many things begin to fall through the cracks.

Suppliers and contractors who receive such document packages can be frightened away from bidding or just as likely will ignore the whole mess and quote what they think they should quote, along with a disclaimer that they hope will get them off the hook if anything goes wrong. Usually, their biggest problem is to *find* the scope of work so that they can decide what the documents really intend to say.

This is not to say that standardization cannot work, only that the question of overkill should be addressed.

In putting together contract bid packages, quite often there is a temptation to assemble an enormous quantity of drawings and data. Certainly, the contractor should have copies of the essential drawings for bidding, but often there are scores of others that he may need for reference. For example, if he is bidding on the mechanical installation of equipment, you may wish him to have access to vendor drawings. As these may be too bulky or numerous to print, a suitable alternative is to list them in the contract with the statement that he has access to them in your office, should he wish to use them.

TYPICAL SPECIFICATIONS FORMAT

A table of contents for specifications may be as follows. Packaging and freight requirements should be covered in the overall procurement package.

1.0   Introduction
2.0   Scope of work
3.0   Service conditions
4.0   Design criteria
5.0   Specifications
6.0   References
7.0   Guarantees
8.0   Drawings and data requirements
9.0   Schedule requirements
10.0   Form of proposal

A description of each section is given below, followed by a typical example of a specification.

1.0   INTRODUCTION   A single paragraph should suffice, giving the project name, location, and what the specification is for.

2.0   SCOPE OF WORK   This is the one we warned you about. Scope of work should be subdivided into "Work Included" and

"Work Not Included." Do not be wishy-washy — be firm in what you want. Make a list in point form.

3.0 SERVICE CONDITIONS This section should state climatic conditions, whether the equipment is going to be operated indoors or outdoors, and whether it is going to be operated twenty-four hours a day or intermittently. Contaminants such as dust, corrosive atmosphere, heat, or anything that could affect the operation should be spelled out.

4.0 DESIGN CRITERIA Any service factors deduced from 3.0 above should be mentioned. Also, the normal and excessive loads and outside influences should be stated.

5.0 SPECIFICATIONS There are two basic approaches to specifying equipment. One is called a performance specification, where the supplier is expected to select equipment to meet operating criteria. The other is used in situations where equipment is specified in detail and the supplier is expected to confirm that the engineer's selections are correct. In the former case, this section should simply make a statement emphasizing the bidder's responsibility for selection. In the latter case, the intent of the engineer should be stated with a request for confirmation, followed by detailed requirements.

6.0 REFERENCES Drawings, publications, data sheets, and calculations should be listed or attached for reference.

7.0 GUARANTEES Guarantees or warranties sought should be stated. It is extremely important that the start of the guarantee period be defined as, for example, "after delivery," or "after installation and acceptance."

8.0 DRAWINGS AND DATA REQUIREMENTS Most companies use a standard form that shows the quantities and details of vendor drawings, catalogues, spare parts lists, operating manuals, installation manuals, and maintenance manuals that are required.

9.0 SCHEDULE REQUIREMENTS Bid periods should be defined in the request for bids. Schedules for drawings and data requirements should be shown in Section 8.0. This section should simply state field delivery requirements.

10.0 FORM OF PROPOSAL Specific data and format requested with the bid should be listed to make your life easier in bid analysis. If you try to get too detailed in this section, most bidders will ignore half the requirements and you will have to chase them by telephone.

EXAMPLE

As an example of a typical specification, vibrating screens are selected.

1.0   Introduction

This specification covers the supply of vibrating screens for the Loose Tooth Bone Marrow Crushing Plant, located in Fuzznuts, Wyoming.

2.0   Scope of Work

2.1   WORK INCLUDED
- Supply of two (2) only 4 feet × 10 feet single deck vibrating screens, complete with screen cloth, V-belt drives, and motors.

  Equipment No. 279 – V – 1
  279 – V – 2

- Supply of one (1) double deck circular vibrating screen, of nominal size 4 feet in diameter, complete with screen cloths, V–belt drives, and motors.

  Equipment No. 299 – V – 1

2.2   WORK NOT INCLUDED
- Support steel
- Vibration isolators
- Feed and discharge chutes
- Dust collection ducting

3.0   Service Conditions

Screens will operate indoors in a building heated to 50°F in winter, but inside temperatures should reach 100°F during summer months. Screens will operate sixteen hours per day five days per week. Ambient dust concentrations are expected to be heavy.

4.0   Design Criteria

Screens shall be designed for twenty-four-hour service, seven days per week, even though actual working hours may be shorter.

4.1 EQUIPMENT NOS. 249 – V – 1
            149 – V – 2

Each screen shall be designed to scalp off 85 percent plus one-quarter-inch material at a feed rate of 20 TPH. Distribution of feed is as follows:

+ 1/4-inch	20%
– 1/4 + 10M	35%
– 10M + 35M	24%
– 35M + 200M	12%
– 200M	8%
	100%

Bulk density of feed = 75 lb/cu. ft.

4.2 EQUIPMENT NO. 299 – V – 1   Screen shall be fed at a rate of 10 TPH by screw feeder. Top deck shall remove 95 percent plus 10M material. Bottom deck shall remove 65 percent + 100M material. Size distribution of feed as follows:

+ 10M	10.2%
– 10M + 20M	33.4%
– 20M + 65M	24.2%
– 100M	9.5%
– 65M + 100M	22.7%
	100%

Bulk density 35 lb/cu ft.

5.0 Specifications

5.1 Frames shall be of 304SS construction.

5.2 Decks shall be 316SS wire cloth.

5.3 V-belt drives shall have a service factor of 2.0 and shall have quick-connect sheaves and 3V-5V-8V series matched belts.

5.4 Bearing shall be spherical roller bearings with taconite seals.

6.0 References

Drawing nos. 5512-M-2 and 512-M-3, attached, show the feed and discharge arrangements as well as mounting of the screen.

7.0   Guarantees

The manufacturer shall guarantee the capacity of the screens and shall warrant them to be free of manufacturing defects for a period of one year after installation or eighteen months after delivery, whichever comes first.

8.0   Drawings and Data Requirements

Seller shall provide drawings and data as per the attached Drawings and Data Requirements Sheet, Form 5932-D.

9.0   Schedule Requirements

Equipment nos. 279-V-1, 279-V-2 are required on site by March 17, 1989. Equipment no. 299-V-1 is required on site by May 10, 1989.

10.0   Form of Proposal

Bidder is requested to provide the following information in his proposal:

Proposal no.	_____
Date of proposal	_____
Unit prices, FOB factory:	
279 – V – 1	_____
279 – V – 2	_____
299 – V – 1	_____
Total price, FOB factory	_____
Taxes	_____
Freight cost	_____
Shipping weight	_____
Delivery point	_____
Delivery time	_____

The above example is typical, although you may elect to go into more detail. Note that it is not a performance specification, even though the supplier has been asked to confirm the sizes selected. Had it been a performance specification, then section 5.0, Specifications, would have been redundant.

CONTRACTS

Typical contracts follow the same general rules laid out for specifications. However, the format will change somewhat, depending on the type of contract.

Even though the contract may be lump sum or cost plus fee, the heart of the contract still remains a good definition of scope of work, which should be placed front and center of the document package. Often scope of work is defined by drawings. These, too, should clearly delineate what is included in the contract and what is not included.

A typical contract package consists of the following:

- Invitation to bid
- Scope of work
- Form of proposal
- Contract agreement
- General conditions
- Specifications
- Drawings and data

A typical contract is too lengthy to provide an example here. However, the following is a check list of items that should be covered or considered in a contract:

- Withdrawl of proposals
- Rejection of proposals
- Performance bonds
- Notices to proceed
- Location of work
- Drawings
- Site inspection
- Labor practices
- Schedules
- Subcontractors
- Exceptions
- Corporate seals
- Signing officers
- Assignment
- Standard specifications
- Lines and grades
- Warranties

- Inspection
- Owner-furnished material
- Contractor-furnished material
- Titles
- Access
- Progress reporting
- Overtime
- Progress payments
- Acceptance
- Holdbacks
- Changes
- Extra work
- Delays
- Termination of contract
- Insurance
- Laws and regulations
- Patent and regulations
- Patent indemnity
- Liens and claims
- Waivers
- Taxes
- Notices of delivery
- Safety
- Sanitation
- Cleanup
- Fire protection and prevention
- Dangerous materials
- Lighting
- Roads and access
- Publicity
- Transportation facilities
- Temporary construction services
- Utilities
- Substitutions
- Measurement and payment
- Codes and standards
- Welding
- Finish
- Inspection, testing and repairs

BREVITY

In preparing specifications and contracts, brevity will help you to get documentation read and complied with. That is not to say that thoroughness should be sacrificed. Indeed, using check lists and putting things in point form can make it easier to be thorough, and the likelihood of facts and instructions being lost in rhetoric is greatly reduced.

CHAPTER

13

CONSTRUCTION MANAGEMENT

An engineering or design manager will argue that the function of construction is to carry out the instructions laid down by the engineering office. The construction manager, on the other hand, is sure to say that the engineering office is simply providing a service to construction, which, after all, is peforming the essential work of the project, i.e., building the plant. In actual fact, neither is entirely correct.

When a play is produced in the theater, the producer assigns the director the task of preparing the play for production. When his work is done, the director formally turns the play over to the production manager, who sees that the performances take place. Project management works the same way, except that the transition from engineering to construction requires a fair amount of overlap and communication, and there is generally never a formal handover. As a result, there is invariably some friction.

PACKAGING

Construction projects can be managed by direct hiring of labor (often through union halls), by breaking the project down into contracts, or by a combination of the two. If you hire an organization to do this for you, they are known as general contractors. It is common to refer to them as "the general."

Many operating companies prefer to act as their own general because they do enough projects to develop in-house construction management expertise, and they like to work with a known quantity. There is a danger, however, of your own people losing competitive sharpness, and your own people are much more difficult to get rid of if things don't work out. And finally, a contractor can be a buffer against litigation.

It is rare, even in a declared direct-hire situation, to avoid using contractors. There are simply too many specialized tasks that require the services of experts. Individuals with very specialized skills just can't be found hanging around a union hall waiting for a call. But one advantage of going direct hire is that it makes it easier to prevent rival unions from winding up at site at the same time. While contractors must declare which unions they employ, surprises are always a possibility.

When it comes to splitting the project into contracts, the real construction management skills come into play. It would seem obvious that it is best to employ as few contractors as possible, simply to make the construction manager's job that much easier — he has a smaller herd to tend. That does not necessarily reduce the number of contractors on site, however, since subcontracts can sprout in multitudes from each main contract let. You may argue that it is not the construction manager's job to ride herd on the subcontractors. If you do, you could be headed for big trouble.

It is a mistake to assume that a contractor only subcontracts to other than his main skill. A good contractor does not hesitate to sub out to another company if he can see an advantage in doing so, or if he thinks he may run short of his own people. This can be especially true if there is another, more lucrative, contract that he can land in the wind.

It is a good idea when starting to plan your project to make a list of work items so that you can get an idea of how you wish to package the job, regardless of whether or not you intend to go direct hire. The following list covers many tasks that may be needed on a typical project:

- Earthwork equipment operation
- Formwork carpentry
- Rebar installation
- Concrete pouring, puddling
- Concrete finishing
- Grouting
- Buried piping and drainage installation
- Earthing installation
- Structural steel erection
- Bricklaying
- Masonry
- Roofing installation
- Siding installation
- Mechanical equipment installation
- Substation
- Electrical power
- Lighting
- Piping
- Ducting
- Heating, ventilation and air conditioning
- Plumbing
- Painting
- Insulation
- Field–erected tanks
- Tank installation
- Culverts installation
- Roof drainage
- Flooring
- Architectural finishing
- Landscaping
- Paving
- Commissioning
- Start up

There could be many more, of course, and there always seem to be specialties added at the eleventh hour.

If your aim is to minimize the number of contracts and subcontracts, then you will consider grouping some of the work items as follows:

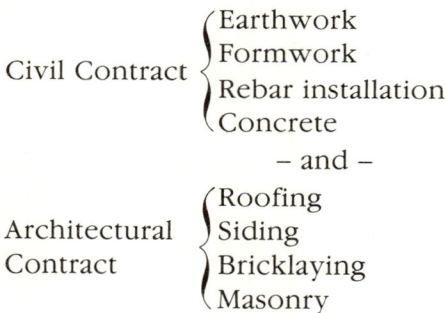

Civil Contract
- Earthwork
- Formwork
- Rebar installation
- Concrete

– and –

Architectural Contract
- Roofing
- Siding
- Bricklaying
- Masonry

There are many possible combinations. However, if you really intend to control subcontracting, you have to make sure during the bidding stage that your contractors can handle the work.

If you intend to allow subcontracting, you should ask the question: Have the contractor and subcontractor worked together before, and if so, what was the track record? It may seem that we are getting ahead of ourselves by delving into this kind of detail, but it is essential that your packaging take into account the abilities of local contractors. Unless your main objective is to offset your role in construction management, there is no point in breaking down your project into, say, five main contracts if there are still going to be twenty-five contractors on the job.

SETTING UP CONTRACTS

Four basic types of contracts are commonly used. These are:

- Fixed price
- Unit price
- Cost-plus-fee
- Bonus/penalty

In general, a fixed-price contract is used when there is a well-defined scope of work and engineering is well advanced or complete. Unit-price and cost-plus-fee contracts are exercised when the scope of work is relatively vague but it is desirable to get the work started. It is assumed that the benefits of an early start outweigh the risks of letting out an ill-defined contract. When capital cost is of prime consideration, a bonus/penalty contract is sometimes let to entice the builder to finish below the estimated cost.

The hard reality of fixed-price contracts is that the owner chooses this method so that he will not experience overruns, but in practice it is extremely rare that a fixed-price contract will not incur extra costs. When a contractor submits his price, he must balance his desire to get the contract with the risk of underbidding. His margin may be wide in good times, but more often than not, it is slim in his eyes. In the execution of the contract, therefore, he is apt to start looking for extras and changes in scope from day one, not just to protect his profit but to guarantee that he will not lose money. It does not matter whether the scope of work comes with a fine breakdown or a broad one — he will use either to advantage. A fine breakdown will force him to work to rule, and any hiccup on the owner's side will incur extras. A broad breakdown, on the other hand, will open the contract to interpretation. A hedge against either of these is to obtain unit prices or cost-plus-fee prices for extras at the bidding stage. At least then you can have some protection against overpricing on extra work.

Unit-price contracts have the disadvantage of requiring thorough monitoring and leave the owner with the appalling feeling that the all-in cost is unknown, even though an estimate has been made. And despite the fact that unit prices have been quoted, there is still the possibility of cost escalation for scope changes. If, for example, you discover that high strength concrete is required in a section of the plant and no such unit prices have been allowed, then the contractor is likely to charge premium prices for this extra.

As an alternative to unit price, sometimes a cost-plus-fee is preferred, even though it suffers from the same vagueness about all-in cost as a unit-price contract. The fee is normally fixed for the entire contract, regardless of the final cost, which should encourage the contractor to perform the work in the shortest time at optimum cost. Quality can suffer if times are good, however, if the contractor tries to spread his men thinly. When times are lean, he may do the opposite — load the job with personnel who would otherwise be idle without pay.

A bonus/penalty contract is based on a fixed price and designed to reward the contractor for performance and penalize him for an overrun. The method carries with it the threat of poor quality work. And the contractor will be, perhaps, even more zealous about seeking extras to protect himself from an overrun than he would be with a standard fixed-price contract.

One trick that has been used regularly is to begin work on a unit-price or cost-plus-fee basis and then convert to lump-sum or fixed-price once the scope of work has been firmly established and understood by both parties. At this point, your estimators should bid against the contractor. This system offers the best and worst of both methods. It does, however, offer the advantage of breathing space at the beginning of the project and allows an update estimate to be made that has an excellent chance of coming in on target. But working with trustworthy, reputable contractors does provide you with the best chance for success and eliminates a good many of the problems outlined above.

SELECTING CONTRACTORS

There is a saying that goes, "The best contractor is the last one who worked for you, and the worst is the one who is presently working for you." If you are fortunate enough to be able to consider contractors with whom you have had direct experience, then you have a leg up on the selection process.

Choosing a contractor begins with a prequalification procedure which should include checking with his previous clients. In government projects there may be no such chance, since public tenders may be called. But in private enterprise you can save yourself a lot of grief by doing a thorough screening of candidates, preselecting only those who can fulfill your needs. It is your chance to make sure that the low bidder can, indeed, do the job and that you will not have to be in the embarrassing position of shooting down a winning bid.

Obviously, your prequalification screening process should concentrate on finding out if the contractor has done similar work. Be wary of the contractor who has only done such work on a joint-venture basis: the partner may have had the real expertise and the bidder may only have provided manpower. Also, if similar experience is not all that recent, the bidder may no longer have the personnel that he once had to manage the job.

Second, check out the contractor's present and future workload. If he is running flat out, he will likely bid high, which could just waste your time. If he badly needs work, chances are he has let his men go. Therefore you must ask the question: Who is going to do the work?

Third, check his financial credibility. Is he going to go bankrupt halfway through your project?

Fourth, what unions does he use and when does the present contract run out? You will not want to see a strike halfway through your project.

Theoretically, if you have done your homework properly, you will have narrowed the field to five or six bidders, and the bids should be fairly close. The selection process should then be primarily on low price.

Unfortunately, theory and practice can sometimes be worlds apart. If you have, say, three similar bids, another that is extremely low, and one that is extremely high, then it is in your interest to find out why there are major discrepancies. Has the high bidder discovered something that everyone else missed? Has the low bidder forgotten something in his bid? In this case, a line-by-line review of the high and the low bids should be made. It is not unusual to ask a low bidder to reconsider his bid if it is substantially low, because your project could suffer in quality if the contractor realizes — too late — that he has underbid.

Once the price factors and scope of work have been reviewed with all bidders who are in the running, then a preaward meeting should be held with the recommended bidder. The purpose of this meeting is to cover *all* of the conditions of the contract on a line-by-line basis. Even if standard contract conditions that are common to the industry have been specified, it makes sense to cover each item one at a time. The odds are that some surprises will surface.

CONTRACT MONITORING

If you have twenty contractors on site and you blithely assume that all twenty are performing to your expectations, you could be in for a rude shock. Even though contracts may be paid with progress payments, one should never rely on the contractor's word that the progress has been made. That is not to imply that contractors are by nature inclined to be unscrupulous, merely that they are people and people are not infallible. And even with the best of intentions, the need to get money up front to finance the cost of a job can very quickly become an unwillingness to actually spend the funds if the temptation to allow the cash to gather interest is too great. Many contractors make it a policy to separate cash in from cash out by the widest possible margin in order to maximize profits.

For contracts on the critical path in particular, you should consider the job as a barrel of apples in a dark room. You have to touch each one to find out if there is a rotten one. Quality and productivity are the two facets to monitor. Quality is checked by inspectors; productivity is measured through time sheets and by quantities surveyors.

A simple example is the laying of bricks. If a contract is to lay 10,000 bricks in one month, then after two weeks, 5,000 bricks should have been laid. An inspector should check to see that the right bricks and mortar were used and that the walls are straight, and a quantities surveyor should count the bricks. If the work was performed by direct hire, then the productivity will be measured by the number of bricks laid per man over a given time period. The cost will be compared to the budget. The same principles apply for laying cable, moving earth, and erecting steel. The key is to set up a means of monitoring on a systematic basis.

Progress Reporting

Weekly and monthly reports should be demanded from each contractor and, in the case of direct hire, from each trade foreman. These, in themselves, become a monitoring tool, but progress reports *should never be considered a substitute for monitoring*. On direct-hire projects the time sheets, inspection reports, and quantities surveyor reports are combined and analyzed by construction management, and action is taken where necessary.

Typical progress reports will have the following headings:

- Contract statement (lump-sum, unit-price)
- Weather report
- Manpower
- Materials
- Budget
- Schedule
- Safety (accident report)
- Problem areas
- Action plan

CHANGE ORDERS

Whether a contract is fixed-priced, unit-price, cost-plus-fee, or bonus/penalty, you can count on having change orders. They are like taxes — aggravating but inevitable. What's more, they will come at you with a sense of urgency. You had better be prepared with an effective procedure, especially if you have twenty or so contractors, all agitating for approvals at the same time.

Any contractor worth his salt will not make substitutions or other changes without authorization if they are going to cost extra, and sometimes even if they save money. Oral instructions may be acceptable at times, but in general, a contractor will want to see authorization in writing.

Change order forms vary, but in essence they should state the following:

- Description of change
- Identification numbering
- Effect on cost
- Effect on schedule
- Justification for change
- Approvals
- Distribution

Distribution is as important as approvals, because how can you control costs, schedule, and quality if concerned parties are not advised? Approvals can be delegated up to certain cost levels, depending on what the project manager feels comfortable with. However, it is usually best to group change orders for approval on a weekly basis.

Field Orders

Field work orders and field purchase orders can be defined as expenditures made without direct input from the engineering and procurement office. As a result, they become a source of resentment and controversy and can create friction between the field and office. In a sense, they are the first step in the process of shifting control of the project from engineering to the site, and that apronstring-cutting exercise can sometimes be painful.

It is usually understood that field purchases and control authorizations are inevitable in a small way. After all, traffic signs have to be bought; construction shacks, johnnys-on-the-spot, unexpected drainage culverts have to be put in. When there is a budget for these items and when the quantities and costs are minor, there is usually no problem. When the effort becomes intensified, however, due to pressures and events at the site, the engineering team see their budget threatened, and the home office procurement team see their authority being usurped. And the sparks begin to fly.

Field work orders that are authorized to a limit of, say, $5,000 without the project manager's approval can be a pain in the neck for a construction manager who needs to get a job done quickly. If the job looks like it will exceed this amount, an old trick is to break the job down into parts, each of which is less than the authorized amount. Then each part is submitted as an individual field order. Another method is to award the order based on an estimate of $4,999 and then say "Whoops! Sorry!" when the actual cost comes in at $25,000.

Naturally, the project manager too is in a bit of a fix. If he insists on approving all of these orders, then he will find that 1) they are presented to him as being very urgent, and 2) if he approves, his engineering and procurement team will feel as though they have been bypassed. For, after all, what is a field work order but an expeditious means of getting things done without having to go through the bureaucratic chain of approvals at head office? As usual, if the project manager insists on approving them, then he has no choice but to face the wrath of his engineering and procurement people.

But who said his job is easy?

SAFETY

Unfortunately, the sign outside a site that reads "XX days worked without a lost-time accident" has become a somewhat pathetic attempt to draw attention to safety, and although it may be the first thing that employees see when they come onto the site, it may appear to be of questionable value.

Concern for safety on a construction project should be inspired entirely by concern for our fellow men. If that is not enough, however, then there are plenty of good sound business reasons for promoting safety, and if you run into budgetary restraints, your construction safety association can provide plenty of back-up cost data.

Promoting safety on the job is a full-time responsibility, and your safety manager will try to reinforce awareness, for example, by erecting the sign mentioned above. He should also host regular meetings with foremen and labor relations men and provide prizes and other incentives for workers and tradesmen. And last but not least, he should be granted the authority to enforce safe work practices on the job. Too often the safety officer is like a guard dog with no teeth, which is surprising, considering that ultimate responsibility — by law — rests with the project manager.

UNIONS

A technician who represented a manufacturer of hydraulic equipment realized that a system needed minor adjustment — just a half-turn of a screw. He went to find a ladder, then propped it up, ready to ascend to the equipment. He had one foot on the first rung of the ladder when a shop steward tapped him on the shoulder. The technician was required to find a foreman, who got two laborers to return the ladder to where he found it. Then the same two carried it back to the equipment for him.

Shaking his head in disbelief, our technician climbed the ladder. After studying the problem, he removed a screwdriver from his pocket. Just as he was about to make the adjustment, he heard a shrill whistle. You guessed it — it was another shop steward. This time he had to contact another foreman, who provided a mechanic to turn the screw plus a helper to hold the ladder!

On another project where productivity had been particularly low, two union strong-men took a feisty foreman behind the warehouse and broke both his arms. "Next time," they warned, "it'll be your legs, too!"

On a major nuclear project a carpenter built a sleeping cabin for himself in the bowels of the formwork so that he could rest from a second job that he had at night. On that same project, the foreman was known to chastise his crew for moving too fast. "You don't want to get laid off, do you?"

How did we get into this mess?

In a major electrical hardware manufacturing plant in the Midwest, management had all the windows painted black so that workers would not gaze out. They had all the doors taken off the toilet stalls and placed a guard in the washroom to report on workers who were too slow in relieving themselves.

If management looks upon labor with disrespect and labor considers management "them," then relations tend to deteriorate with time. Labor relations officers sometimes feel they must handle problems by fielding grievances instead of preventing them from developing in the first place. Just orchestrating unions so that one does not step on the toes of the other is also not the final solution.

Weekly, if not daily, contact should be encouraged — not just between labor relations, personnel, and the stewards, but also between the workers and the management team. If there is a dispute with a foreman or engineer and a tradesman, for example, the steward, the tradesman, the offending party *and* a management representative should sort it out together.

Major unions usually associated with construction projects are:

- International Association of Heat and Frost Insulators and Asbestos Workers
- International Brotherhood of Boilermakers, Iron Ship Builders, Blacksmiths, Forgers, and Helpers
- International Union of Bricklayers and Allied Craftsmen
- Union Brotherhood of Carpenters and Joiners of America (also covers millwrights)
- International Brotherhood of Electrical Workers
- International Union of Elevator Constructors
- International Union of Operating Engineers
- International Association of Bridge, Structural, and Ornamental Ironworkers
- Laborers International Union of North America
- Tile, Marble, Terrazzo Finishers and Shopmen International Union
- International Brotherhood of Painters and Allied Trades
- Operative Plasterers and Cement Masons International Association of the United States and Canada
- United Union Of Roofers, Waterproofers and Allied Workers
- Sheet Metal Workers International Association
- United Association of Journeymen and Apprentices of the Plumbers and Pipefitting Industry of the US and Canada
- Teamsters Union

Overtime

It has been past practice to schedule overtime in order to recover schedules or meet deadlines. Experience has taught us that this is not economical. By the time overtime pay is added at time-and-a-half, double-time, and triple-time, and the attendant reduction of efficiency is taken into account, it is usually better to simply add more bodies to the work force at straight time.

PUBLIC RELATIONS

There is nothing worse than getting halfway through a project and finding a protest group ready to do battle along the fence line. As often as not, the protest is born of ignorance. A proper job of public relations can prevent this kind of situation.

First off, press releases announcing the project should be sent to the local newspapers long before the bulldozers arrive. Then, a string of releases should be made throughout the life of the project, perhaps focusing on local personalities who are benefactors of the job.

It is also a good idea to hold a public meeting early in the project to explain what it is all about, what benefits there will be, what problems will arise — both short- and long-term — and to answer questions from the public.

CHAPTER

14

REPORTING

How we tend to report:
>"How're you doing?"
>"Fine, thank you. Everything is perfect."

How we never report:
>"How are you?"
>"Rotten. Got three hours? I'll tell you about it."

Or there were the two psychiatrists who passed on the street.
>"Good morning," said one psychiatrist.
>"Good morning," said the other.
>"I wonder what he meant by that?" they both thought.

MIS (Management Information Systems) is a fancy acronym that could mean Methods of Insight Stimulation but sometimes becomes Muzzling of the Internal Scenario. Reporting is what it is really all about. How do we get a true picture of what is happening up through the organization to the project manager so that he can effectively direct the project? Some people who write reports would rather make the picture rosy — everything is fine, thank you very much—while others write long essays and provide reams of back-up data that say relatively little about the true health of the project.

Information flow to the project manager is from the floor of engineering and project management services and from the site. The project manager, in turn, reports to his superiors, who could be the board of directors.

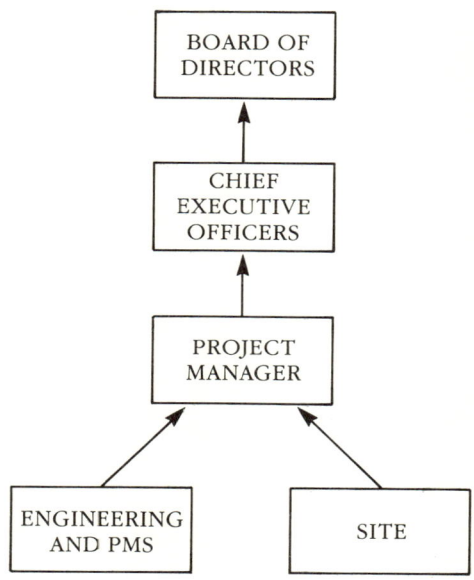

FREQUENCY OF REPORTING

It is customary to report to the board of directors on a quarterly basis, or in the case of calamity and requests for additional funds, as the need arises. Company presidents or chief executive officers generally require monthly reports. Reporting from engineering and site are best done on a monthly basis. However, the maintenance of data must be done on a weekly basis for it to be ready for review by the project manager at any time.

REPORTS TO THE PROJECT MANAGER

Engineering

In order to evaluate progress, the project manager needs to know the status of:

- Manhours
- Drawing production
- Specification and contract writing production
- Vendor drawings and data

He also needs to know if there are problem areas. However, these usually arise at weekly meetings and are identified on a punch list. Some of the data on specifications, contracts, and vendor drawings may overlap with procurement and therefore may be routed from engineering through procurement, depending on the organization and size.

Manhour reporting usually begins with a time sheet that is filled out by each employee. Even if employees are paid monthly or bimonthly, it is best to report time distribution weekly, otherwise the booking of individual time will become unreliable due to short memories and the longer time interval between reports. Time sheets are submitted to discipline supervisors for checking and logging. Cost coding may be done at this level, provided that the system used is not complex. If it is complex, there will be many inaccuracies. If a complex cost coding system must be used, then it is best to have it coded by a cost clerk.

Once time charges have been coded, logging and comparing with budgets can begin. It should be mentioned that this data will be used with drawings, specifications,and contracts production to measure productivity. Each discipline should log the time on a cost code basis which will identify the work by plant areas and also by general activity, such as drawings or engineering.

It is tempting at this point to require engineers and draftsmen to record their time on an individual specification, contract, drawing, or other package basis. However, in practice this becomes unreasonably cumbersome, especially when the person works on several items each day. The benefits gained are not worth the effort it takes to see that the logging is done properly. One large consulting group still monitors work in this fashion, but the reality is that the logging is fudged to suit the system.

Figure 14.1 is a typical weekly log sheet for an engineering discipline. Data from this is extracted on a monthly basis and recorded on an engineering monthly manhour report and summary (Figures 14.2 and 14.3). This data can be recorded manually or on a computer and can be sorted in different ways.

Drawing progress reports are made for each discipline. Figure

FIG. 14.1 — WEEKLY ENGINEERING MANHOUR LOG

Project No. 5199   Project: Boiling Bones Rendering Plant   Discipline Mechanical

Week Ending July 27, 1989

Cost Code	Description	Manhour Distribution		
		Drawings	Engineering	Total
5199.100	Boiler Plant	120	44	164
5199.200	Bone Crushing Plant	119	39	158
5199.300	Grease Mill	132	57	189
5199.400	Bottling Plant	43	3	46
5199.500	Clarification Pond	17	2	19
	TOTALS, THIS WEEK	431	145	576

14.4 is an example. This information can also be put on computer, but it is unlikely that savings in manpower will result.

Although the procurement people are heavily involved in the purchase of equipment, the control function should rest with engineering until purchase orders are placed. A mechanical equipment list like the one shown in Figure 14.5 is a key in controlling the procurement of equipment. It advises electrical and instrumentation people of requirements, and tells inspection, shipping, and expediting well in advance what is forthcoming. Then the monitoring of the procurement process up to the placement of purchase orders is done on a form similar to the one in Figure 14.6. Contract monitoring to the award stage is handled with similar documentation.

Manpower planning and levelling is monitored on a form like the one in Figure 14.7, which shows how your staff build-up compares with your estimate. This can serve as an indicator of productivity.

In general, productivity is measured in manhours per drawing or specification. Since production of drawings depends on engineering and procurement activities, some companies use drawing production as the sole measure of productivity. Certainly, if manhours spent are compared to percentage of completed drawings — or specifications and contracts, for that matter — a measure of how the project is doing as compared to expectations originating with the budget estimate can be made. For example, if 180 drawings have been estimated to take an average of 150 manhours each (180 × 150 = 27,000 hours), you can measure the weighted percentage completed from the drawing progress report (Figure 14.4) to get an idea of your productivity. If you have 160 drawings that are, on average, 60 percent complete, then you should have expended .60 × 160 × 150 = 14,400 manhours. If you have really used 15,000 drafting manhours, then your productivity is

$$\frac{14,400}{15,500} \times 100 = 96 \text{ percent.}$$

Control of vendor prints has been discussed in Chapter 11. Since these are an essential part of the engineering chain, they should be reported to the project manager in a status report (Figure 14.8).

FIG. 14.2 — MONTHLY ENGINEERING MANHOUR REPORT

Project No.   5199

Cost Code	Description	This Month	Total To Date
5199.100	Boiler Plant	480	5085
5199.200	Bone Crushing	449	4470
5199.300	Grease Mill	332	3711
5199.400	Bottling Plant	163	1600
5199.500	Clarification Pond	60	715
	SUB-TOTALS, DRAWINGS	1484	15581
5199.100	Boiler Plant	161	1577
5199.200	Bone Crushing	159	1589
5199.300	Grease Mill	243	2313
5199.400	Bottling Plant	12	121
5199.500	Clarification Pond	10	99
	SUB-TOTAL, ENGINEERING	585	5699
	TOTAL — MECHANICAL	2069	21280

Project: Boiling Bones Rendering Plant

Discipline Mechanical
Month: _July '89_

Drawing Manhours			
Budget	To Complete	Forecast To Complete	Forecast Over (under)
11000	5915	7500	1585
12000	7530	7530	—
9000	5289	5289	—
5000	3400	3000	(400)
2500	1785	1785	—
39500	23919	25104	1185

Engineering Manhours			
3000	1423	1800	377
3000	1411	1411	—
4500	2187	2187	—
500	379	500	121
400	301	301	—
11400	5701	6199	498
50900	29620	31303	1683

FIG. 14.3 — MONTHLY ENGINEERING MANHOUR
SUMMARY

Project No. 5199

Discipline		This Month	Total To Date
Mechanical	Drawings	1484	15581
	Engineering	585	5699
	Sub-total	2069	21280
Civil	Drawings	1979	21100
	Engineering	844	8007
	Sub-total	2823	29107
Architectural	Drawings	211	2001
	Engineering	27	295
	Sub-total	238	2296
Electrical	Drawings	799	7850
	Engineering	560	5500
	Sub-total	1359	13350
Instrumentation	Drawings	120	1010
	Engineering	119	1201
	Sub-total	239	2211
	TOTAL	6728	68244

Project: **Boiling Bones Rendering Plant**   **Month:** _July '89_

Budget	To Complete	Forecast To Complete	Forecast Over (under)
39500	23919	25104	1185
11400	5701	6199	498
50900	29620	31303	1683
45000	23900	25000	1100
12000	3993	5000	1007
57000	27893	30000	2107
4500	2499	2200	(299)
1500	1205	1205	—
6000	3704	3405	(299)
20000	12150	12150	—
10000	4500	4500	—
30000	16650	16650	—
7800	6790	6500	(290)
7500	6299	6000	(299)
15300	13089	12500	(589)
159200	90956	93858	2902

FIG. 14.4 — DRAWING PROGRESS REPORT

Project No.   5199

Progress: To Date ▰▰▰

Project: Boiled Bones
Rendering Plant
This Period ⊐⊔⊔⊔⊏

Drawing Number	Title	Progress					Rev	Date
		20	40	60	80	100		
100-C-1	Plot Plan						A	5-31
100-C-2	Floor Plan EL.343						A	7-21
100-C-3	Sections A-A, B-B						A	7-21
100-C-4	EL.343'Details						A	7-21
100-C-5	EL.347'Details						A	7-21
	−et cetera−							

Description: Boiler Plant   **Date:** *July 3/1989*
Cost Code: 5199.100
Discipline: Civil

Issue — Letters: before cons't 0 = issued for cons't							Cons't Issue Sch	Act	Remarks
B 6-15	C 6-30	O 7-10	I 7-15				7-10	7-10	
							7-31		*Issued for bids contract C-101*
							7-31		//
							7-31		//
							7-31		//

FIG. 14.5 — MECHANICAL EQUIPMENT LIST

Project No. 5199

Equipment Number	Qty	Description	Cost Code
100-T-01 thru -03	3	Sludge Tanks	100.1237
100-P-01 thru -07	7	Sludge Pumps	100.1239

Project: Boiling Bones
Rendering Plant

Date: July 31, 1989
Revision 4

Spec. No.	Nominal Capacity	Size	HP × RPM	Weight, each lb.
100-M-01	1600 gal.	4'0 × 3'	–	1500 lb.
100-M-15	75 US GPM	3 × 5	7.5@ 1800	212 lb.

FIG. 14.6 — EQUIPMENT PROCESSED FOR PURCHASE

Project No. 5199

Project: Boiling Bones Rendering Plant
A = Actual
S = Scheduled

Spec. No.	Rev.	Description		Spec. App'd	Out for Bids	Bids Rec'd
100-E-01	0	Wound rotor motors	A	3-1	3-15	4-15
			S	3-10	3-15	4-12
100-E-02	0	Switchgear	A	3-1	3-15	4-15
			S	3-1	3-15	4-20
100-E-03	1	Transformers	A	6-10	6-15	7-15
			S	6-10	6-20	7-31

Date: July 31/89
Discipline: Electrical

Recom. for Purch	Purch Auth.	P.O. Issued	Progress				Remarks
			25	50	75	100	
5-15	6-30	7-15	███	███	███		Awarded to G.E
5-16	6-30	7-20					
4-30	5-15	5-20	███	███	███		Awarded to Sq. D.
4-30	5-12	6-30					
7-31	8-15	8-31	███				

FIG. 14.7 — ENGINEERING MANPOWER PLANNING

Month No.		1	2	3	4	5	6	7	8	9	10	11
Mech	B	12	24	24	24	24	34	34	44	44	44	44
	A	10	18	24	26	26	32	34	44	44		
Civil	B	7	10	10	10	25	28	28	32	42	62	62
	A	7	10	10	10	25	28	28	28	40		
Arch	B						2	2	2	2	2	2
	A						2	2	2	2		
Elect	B	1	3	3	3	8	12	12	16	16	26	26
	A	1	3	3	3	5	12	12	16	16		
Inst	B		3	3	3	3	4	4	4	6		
Total	B	20	40	40	40	60	80	80	100	110	140	140
	A	18	34	40	42	59	78	80	94	108		
Total	B	20	60	100	140	200	280	360	460	570	710	850
Cum.	A	18	52	92	134	193	271	351	445	553		
DIFF		(2)	(8)	(8)	(6)	(7)	(9)	(9)	(15)	(17)		

Week No. 9
Date: July /89

12	13	14	15	16	17	18	19	20	21	22

Cum. Manmonths — 2000, 1500, 1000, 500

12	13	14	15	16	17	18	19	20	21	22
44	44	44	44	34	34	34	34	24	9	9
62	62	62	62	52	52	45	45	25	1	1
2	2	2	2	2	2	2	2	2		
26	26	26	26	16	16	23	23	23	6	6
140	140	140	140	110	110	110	110	70	20	20
990	1130	1270	1410	1520	1630	1740	1850	1920	1940	1960

FIG. 14.8 — MONTHLY VENDOR DATA STATUS REPORT

Project No.: 5199

Project: Boiling Bones Rendering Plant

Discipline	No. of P.O.s	
	Planned	Issued
Mechanical	41	30
Electrical	26	16
Civil	12	12
Architectural	4	4
Instrumentation	19	2
TOTAL	102	64
% COMPLETE		62.7%

Project Management Services

Project management services include all procurement activities, estimating, planning and scheduling, cost control, and administration.

Figure 14.6 pretty well covers the status of equipment being processed for purchase, and a similar form — also provided by the engineering disciplines — should cover contract activities up to the award stage. Reporting from procurement should cover postaward activities, including expediting, inspection, and shipping. Each of these activities must keep individual logs and reports for each purchase order. However, the reporting of such to the project manager should be distilled into a comprehensive form that gives the overall status yet highlights problem areas. Figure 14.9, a procurement status report, is an example of the kind of reporting required.

Although main milestone estimates are the activities that attract the most attention, the estimator's job can be ongoing. He will provide tender check estimates on contracts, perhaps perform trend forecasting, and will continually monitor project development in anticipation of the need for an update estimate. Monthly reporting to the project manager usually is in narrative form, ad-

Date: July '89

No. of Vendor Prints Processed		
Expected	**To Date**	
	Initial	**Final**
457	212	165
220	76	–
1110	630	475
12	–	–
1400	–	–
3189	918	640
	28.8%	20.1%

vising of personnel charging to the job, new data made available, and so on. A master bar chart schedule should be updated on a monthly basis, with periodic CPM analyses commensurate with the complexity and tightness of the schedule included in the estimator's monthly report.

In order to control the cost on his project, the project manager needs to know how much money has been spent, how much has been committed, when future spending is anticipated and at what rate (cash flow), and how all of that stacks up against the respective estimates. It is up to the cost engineer to extract that information from engineering, procurement, and construction and to present it updated on a monthly basis to the project manager. It should be emphasized, however, that the key to cost control is to look at historical information from the point of view of productivity, and from that forecasts need to be made. Look at what is happening now, then look ahead.

The administrative aspect of a project normally does not attract much attention. However, an astute project manager will insist on an accounting of the use of telephone, telex, and photocopy equipment and printing on a regular basis. Although these items tend to get buried in overheads, how can you say you have control if you don't really know what they are?

FIG. 14.9 — PROCUREMENT STATUS REPORT

Project No.   5199

**Project: Boiling Bones
Rendering Plant**

P.O. Number	Equipment Number	Description	Inspection	
			sched.	actual
200-M-01	200-C-01 200-C-02	Bone crusher	6/14/89	6/14/89
200-M-02	200-P-01 200-P-02	Sump pumps	N.R.	
200-M-03	200-VS-01	Vibrating screens	5/15/89	5/15/89

Date: July 31, 1989
Discipline: Mechanical

Problem Areas	Delivery		Via	Remarks
	orig.	latest		
Rotor shaft off spec. Needs metallizing	10/31/89	11/31/89	Rail	Delivery slippage due to rotor shaft off spec.
	9/30/89	9/30/89	Truck	
	10/31/89	10/31/89	Rail	Site not able to accept until 1/15/90. Storage charges may accrue.

Construction

Reporting from the site parallels many of the engineering and project management services reports and in some cases augments or contributes to preparation of such reports from the home office. Contract status reporting is similar to procurement status reporting, as is the field work order status report. Figures 14.10 and 14.11 are examples.

Where productivity, and therefore cost control, is measured by the amount of material used or installed, such as piping or electrical cables, reporting is best done on a weekly basis and then compiled monthly for reporting to the project manager. Figure 14.12 is a typical weekly report for process piping. Bulk materials budget and cost control are similarly reported on a monthly basis, even though recording may be weekly. Figure 14.13 is a typical form.

As forecasting is an essential part of controlling a project, forecasts should be made monthly in the same format as the original estimate, i.e., as per the estimate worksheet in Figure 8.1. Forecasts should be broken down into material, labor, and contract. Figure 14.14 is a typical worksheet.

Input to the master schedule should have been made directly to the planning and scheduling group. Nevertheless, items affecting or threatening the schedule should be included in the field report.

A narrative report should be made on problem areas, safety, and union activities, and any publicity, inadvertent or otherwise, should be mentioned. Also, a list of prominent visitors to the site should be given. And finally, photographs of construction highlights should be made for inclusion in the report.

REPORTING UPWARDS

The project manager is charged with the responsibility of spending large amounts of money on behalf of his president or the chief executive officer or, more simply, on behalf of a vice president or other person who acts as project sponsor. This individual reports to the board.

Depending on the confidence the project sponsor has in his project manager, the monthly report is sufficient to keep him informed. The project manager may present the project sponsor with trend forecasts on a weekly basis, however, if there is a lot of pressure from people outside his jurisdiction to make scope changes. This would be of particular concern if there are a number of investors in the project who send their men in to monitor progress and register individual demands. The project sponsor may then, in turn, elect to transfer the pressure to the board if the budget appears to be threatened.

The Monthly Report

Where are we?

Where are we going?

Quite often, monthly reports spend a lot of time telling readers where you have been. Analysis is confined to justification of what has passed. The importance of the past, however, is secondary to projections of what will happen if things do not change and forecasts of what adjustments need to be made in order to go where you want to go. For example, Figure 14.15 shows an S curve envelope that describes both a projection or trend and a forecasted recovery.

Suggested contents of a monthly report are as follows:

1.0	Introduction
2.0	Summary
3.0	Cost commitment reports
4.0	Schedule
5.0	Engineering progress
6.0	Procurement
7.0	Construction report
8.0	Trends
9.0	Problem areas
10.0	Recovery plans
11.0	Site photographs

1.0 INTRODUCTION   This is a statement of what the report represents, which project it applies to, which month it covers, which report number it is, and to whom it is made.

2.0 SUMMARY   This is simply an abstract of the report, stating generally where you are and where you are going. For example:

"During the month of July, engineering costs and progress remained satisfactory. Construction is approximately one month behind schedule due to adverse weather. Recovery is expected

FIG. 14.10 — CONTRACT STATUS REPORT

Project No.   5199

Project: Boiling Bones Rendering Plant

Contract number	Award date	Title	Contractor	Type of contract	Bid total amount	Latest C.O.#	Current total
100-C-001	3/15/89	Grubbing & clearing	Amalgamated	Unit price	$1,020,000	3	$1,050,000
100-C-002	3/30/89	Foundations	Pourcon	Lump sum	$2,400,000	7	$2,440,000

Date: July 31, 1989

Percent complete	Estimated productivity	Start Date		Completion		Remarks
		sched.	actual	scheduled	forecast	
100%	95%	4/1/89	4/15/89	6/5/89	6/15/89	Completed
79%	101%	6/5/89	6/15/89	8/31/89	8/31/89	

FIG. 14.11 — FIELD WORK ORDER STATUS REPORT

Project No.   5199

Project: Boiling Bones Rendering Plant

F.W.O.#	Description	Award date	Contractor
200F-01	French drain	6/15/89	Frenchy
200F-02	Culvert	6/15/89	Culco
200F-03	Temporary power to area 300	7/15/89	Zappco

Date: July 31, 1989

Estimated cost	% Comp.	Final cost	Remarks
$2,500	50%		
4,500	100%	$4,500	
2,200	20%		

FIG. 14.12 — WEEKLY LABOR REPORT

Project No.   5199

Project: Boiling Bones Rendering Plant

Item	Unit	Quantity			
Process Piping Area 200		**Budget**	**Fore-cast**	**To date**	**This week**
2"&under-fab & install socket w.	LF	18,000			
screwed		6,800			
copper & PVC		2,200			
Subtotal 2" &under		27,000			
Special pipe 3 -10" C.I. mech it.	LF	550			
16-20" C.I. flanged		150			
3 -8" PVC		870			
Subtotal special		1,570			

Sheet __ of __
Week ending: July 31/89

Manhours				Unit manhours			
Budget	Fore-cast	To date	This week	Budget	Fore-cast	To date	This week
18,000				1.00			
3,100				.46			
1,100				.50			
22,200				.82			
330				.60			
150				1.00			
435				.50			
915				.58			

FIG. 14.13 — MATERIAL COST CONTROL

Project No. 5199

Project: Boiling Bones Rendering Plant

Cost code	Description	Supplier	Unit
220.0601	shop-fabricated pipe main steam	Pipco	LF
.0602	condensate		LF
.0672	16 × 14 × 16 3-way valve		EA

Budget			Commitment			
Quantity	Unit price	Total	Quantity	Unit price	Total	Difference
300	75	22,500				
350	40	17,000				
2	15,000	30,000				

FIG. 14.14 — FORECAST WORKSHEET

Project No.   5199

Project: Boiling Bones Rendering Plant

Cost Code	Description	U N I T	Quantity			Direct manhours		
			Budget	To date	Fore-cast	Unit rate	Budget	To date

Sheet __ of __
Date: July 31/89

Fore-cast	Direct labor cost			Contract cost			Total		
	Budget	To date	Fore-cast	Budget	To date	Fore-cast	Budget	To date	Fore-cast

FIG. 14-15 — S-CURVE ENVELOPE OF EXPENDITURES

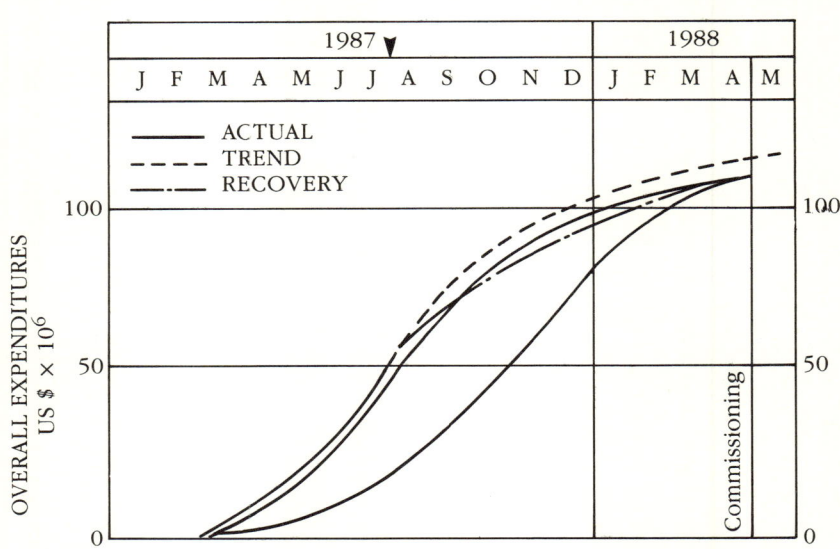

during the next three months, however, by working a six-day week."

3.0 COST AND COMMITMENT REPORTS   S curve envelopes are given for costs and commitments. Sometimes these are categorized by engineering and construction and by purchase and contracts. Curves should show trends and forecasts and be backed up by tables showing budgets and forecasts.

4.0 SCHEDULE   A summary bar chart schedule is given with a narrative discussion of its implications. Also, a milestone schedule should be listed.

5.0 ENGINEERING PROGRESS   Include tables and discussions showing the status of:

- Drawings
- Specifications
- Contracts
- Design
- Manpower

Normally, drawing progress is summarized from the drawing progress report (Figure 14.4), and calculated productivity is shown.

Specifications and contract progress are similarly summarized. Design progress is reported narratively. Manpower is reported by inclusion of a manpower planning graph (Figure 14.7), supported by a narrative that highlights under- or overrunning.

6.0 PROCUREMENT   In general, purchasing has been covered by the engineering section. However, a list of purchase orders and contracts and the amount committed to them should be listed. Then a narrative describing other procurement activities is required.

7.0 CONSTRUCTION   Manhours, material, and contract expenditures should be tabled with a calculation of productivity shown. These should be followed by a narrative highlighting monthly activities at the site.

8.0 TRENDS   Trends in costs and schedules of engineering, procurement, and construction should be noted, referring to cost and commitment curves as well as productivity. Major trend forecasts should be mentioned, and there should be a discussion of their frequency and tendency to become change orders.

9.0 PROBLEM AREAS   Focus on problem areas in engineering, procurement, and construction. It's the chance for the project manager to whine!

10.0 RECOVERY PLANS   Having finished crying, the project manager must now put forth his plan for recovery. If, for example, the cost of installing electrical cable has proven more than expected, what is he going to do about it? Reduce scope? Fire the contractor? Live with it?

11.0 SITE PHOTOGRAPHS   This section puts a touch of reality into the report. Photographs always attract interest and are very satisfying, unless they show such things as a crane that has fallen over or an embankment that has slumped. But these too are a great aid in describing site problems.

Trend Forecasts

As discussed in detail in Chapter 10, trend forecasts are an early warning system for identifying probable changes in scope. They also give the project manager some control over outside influences on his project. If the tendency is for trends to be rampant, then this should be a signal to the project manager that an overall scope review and perhaps a new estimate should take place. At that point, an analysis and special report, with recommendations, should be projected upwards through the project sponsor to the board.

Keeping It from the Old Man

Once in a while, a project sponsor becomes overbearing, placing unrealistic demands on a project. Even with trend forecasts to protect him, the project manager occasionally finds that he must keep things from The Old Man in order to maintain steady progress. This is an extremely dangerous maneuver, especially if there is a chance of things not working out. It should be avoided if possible. It is also risky to attempt to go over the project sponsor's head. He may find out.

One major project was being controlled by a steering committee instead of a single project sponsor. Although there were one or two competent individuals among them, the members of the committee were constantly in disagreement and were very reluctant to make decisions. Frequently, they turned the blame for problems on the project manager. They became known on the project as the "broken rudder committee." With schedule slipping and costs escalating and after strong pleas were made through trend forecasting, the exasperated project manager finally went over the heads of the committee to the president of the company. He tried to have the committee pared down to one or two decisive individuals.

"That's not possible," said the president. "They're a committee because each of them alone is incompetent."

At least, thought the project manager, someone sympathizes and understands.

CHAPTER
15

ADMINISTRATION

Many years ago, an engineer joined a small consulting company to replace their only mechanical engineer, who had been fired, presumably because he could not get along with a project manager. The company had a contract to build a cement plant and had concentrated its efforts on building the concrete product storage silos and on the materials handling that went with them because the owner wanted to begin selling in the area well before the plant got finished. He could easily fill the silos from his other plants. This strategy would allow the plant to come up to full production immediately upon completion.

Much to the engineer's chagrin, his predecessor had left only a few pages of calculations in his desk. It would be necessary, thought the engineer, to go through the central files to properly acquaint himself with the project. The project manager was a very accommodating fellow and immediately showed him where the project files were kept. "The mechanical files are in here," he said, pointing to the second drawer of a four-drawer filing cabinet.

In it was a single folder marked "Mechanical." It contained a quotation on a dust collector, a contract for air slides, and some brochures. This was all, for a multimillion dollar project that was 25 percent complete?!!

Now, it would have been easy to lay the blame for that slim volume of files on the mechanical engineer who had vacated the post. After investigation, however, it was found that other branches of the engineering profession were similarly ill represented. So was procurement, construction, and any other skill one might consider essential to running a project. Of course, this situation could not be permitted to continue, and appropriate action was taken. The point of this story is to illustrate the general attitude of most technical people towards administration, and that is, "Someone else will take care of it." In this case, that "someone else" had to be the new engineer.

Administration is a sort of foggy catch word that covers all of the mundane details of running an office: timekeeping, filing, mailing, distribution, printing, and drawing control, to name a few. Systems for running an office vary from simple chronological filing to complex, computer-integrated behemoths that are staggering in the amount of detail they include. MIS — Management Information Systems — is a popular acronym, a buzzword that is haunting the office crowds these days, and it takes the likes of a systems analyst to fully understand what it means.

Management of capital projects, however, uses repeatable, logical systems that should not evolve into overly complex administrative systems as often has happened in the past. The KISS principle — Keep It Simple, Stupid — is paramount in dealing with technical people, primarily because they tend to have tunnel vision and get totally involved in the technical problems at hand. The less attention they have to pay to administrative matters, the better. Your administrative systems, then, should optimize the bureaucratic process, and an analyst should only be required if you are converting to a computerized system. If you try to impose a complex, time-consuming system on your technical staff, the quality of their work and your system will suffer. Provide "someone else" to do it if you want it done right.

In order to set up for a capital project, realize that administration is a service to the project. Make a list of who or what is going to be served by administration, referring to an organization chart. At least the following will usually be included in your list:

– Management
– Engineering
– Procurement
– Estimating
– Accounting
– Planning and Scheduling
– Construction
– Payroll

Next, make a list of what each of the above produces or does for the project. Keep this list open: You will add to it as you think of other things or as items come up.

Management — Gives instructions to project.
Reports on projects to client or owner or board.
Approves cost estimates.
Approves schedules.
Approves engineering.
Approves purchases.
Approves contracts.
Staffs projects.
Monitors cost control.

Engineering — Designs plant.
Writes specifications.
Writes contracts.
Reviews, analyses bids.
Recommends or approves purchases.
Reviews vendor drawings.
Recommends or approves contracts awards.
Reports to management on project progress.
Makes/approves estimates.
Staffs project.
Monitors cost control.
Makes/approves schedules.
Coordinates with construction.
Approves time charges.

Procurement — Awards purchase orders.
Awards contracts.
Expedites deliveries.
Inspects equipment.

Obtains pricing.
Does commercial evaluations of suppliers.
Makes up bid lists.
Writes general conditions.
Researches suppliers.
Reviews, analyses bids.
Approves time charges.
Staffs project.

Estimating — Prepares capital cost estimate.
Obtains bids/pricing.
Approves time charges.
Prepares check estimates.
Staffs projects.

Cost Control — Prepares cash flows.
Prepares reconciliations.
Prepares trends.
Reports costs to project.

Accounting — Pays bills.
Prepares reconciliations.
Provides statements.
Monitors bank accounts.
Deposits money.
Keeps daily journals.
Staffs accounting group.

Planning and
Scheduling — Prepares work breakdown.
Prepares networks.
Prepares bar charts.
Monitors schedules and updates them.
Does manpower levelling.

Construction — Supervises construction personnel.
Selects contractors.
Monitors and negotiates changes.
Supervises, organizes safety.
Obtains permits.
Obtains work lists.
Prepares equipment and tools lists.

Obtains consumables.
Prepares schedules.
Controls costs.
Looks after labor relations.
Coordinates work.
Sequences work.
Does manpower levelling.

Payroll — Provides timekeeping.
Prepares paychecks.
Pays payroll taxes.

Administration — Keeps project files.
Provides stenographic services.
Runs printing of drawings.
Provides office supplies.
Provides office equipment.
Runs teletype, telephone, communication
equipment.
Provides computer service.
Provides distribution and handling of mail and
courier.
Maintains a library.

Once you have a good idea of what everyone does, then make a list of what each group needs to do its job and from whom, referring to the above.

	What	From Whom
Management	– Regular reports	– Each group
	– Cost estimates	– Estimating
	– Schedules	– Planning and scheduling
	– Plant design drawings	– Engineering
	– Specifications and contracts, bid requests	– Engineering, procurement, construction
	– Bid analysis and recommendations for purchase	– Engineering, procurement, construction cost control
	– Cost reports	– Cost control

Engineering	– Commercial evaluations	– Procurement
	– Bid lists	– Procurement
	– General conditions	– Procurement
	– Budget approvals	– Management
	– Estimates	– Estimating
	– Schedules	– Planning and scheduling, construction
	– Purchase approvals	– Management procurement
	– Contract approvals	– Management procurement, construction
	– Drawings and specification approval	– Management construction
Procurement	– Requests for bids	– Management, engineering, construction
	– Requests for purchase	– Management, engineering, construction
	– Inspection instructions	– Engineering, construction
	– Expediting orders	– Construction
Estimating	– Drawings, specifications	– Engineering
	– Prices	– Engineering, procurement
	– Contracts	– Engineering, construction
Cost Control	– Estimates	– Estimating
	– Schedules	– Planning and scheduling
	– Pricing	– Procurement
	– Billings	– Accounting
	– Trends	– Trend forecaster

Planning and Scheduling	– Durations	– Procurement, engineering, construction, estimating
Construction	– Drawings, specifications	– Engineering
	– Commercial evaluations	– Procurement
	– Bid lists	– Procurement, engineering
	– Budget approvals	– Management, engineering
	– Purchase/contract approvals	– Management
Payroll	– Time sheets	– Each group

Having organized a rough list of who does what with whom, the administrator of a project now must sit with the project manager to sort out how this should all be controlled. It should be pointed out that in a small project, the project manager's secretary could very well become the administrator. And on a very small project, the project manager himself and his engineers may have to do many of the things that "someone else" should do. Nevertheless, we can assume a reasonably large project for this exercise in order to cover all the bases, taking the administrator's tasks one at a time.

FILING SYSTEMS

Filing systems provide for the storage and retrieval of information. A second important function of filing systems is to provide legal back-up in the event of litigation. Professional engineers have some personal vulnerability, but generally they are under the legal umbrella of the corporations who employ them. Filing systems, therefore, can be regarded as "protect-your-ass" data banks from which you can draw facts to back yourself up when things go wrong.

Ideally, all filing belongs in a central file for use by each group. In practice, however, each group will insist on their own set of files, whether you prohibit it or not, and individuals will tend to squirrel things away for easy reference. It is a good idea, therefore, to create both project files for record purposes (and the exclusive use of the project manager) and working files on the project floor. Electronic filing usually falls into the latter category and provides hard copies to the project files.

Other broad filing classifications in project management are as follows:

- Procurement
- Contracts
- Correspondence
- Interoffice correspondence
- Engineering
- Cost control
- Estimating
- Construction
- Drawings
- Vendor drawings

Procurement files should try to cover the procurement chain for each piece of equipment purchased. A single file should be kept for each piece. Each file should provide the history of that piece of equipment, beginning with requests for bids and ending with a purchase order and its revisions.

Contract files should be treated the same way as procurement files.

Correspondence files are meant to cover all incoming and outgoing correspondence. You may elect to consider correspondence outside the project team to be in this category. Subdividing this file by subject must be done. In addition, a daily log of all incoming mail, regardless of its final destination, should be made, and a copy of all outgoing correspondence should be kept in a chronological file. Telecommunications should be filed by subject, with a copy to an ingoing or outgoing chronological file.

Interoffice correspondence files are used to store memos, minutes of meetings, and other communications between the members of the project teams. Usually this file is a little less formal and is often used on a "protect-your-ass" basis between departments.

Engineering files should contain design briefs and copies of correspondence and interoffice correspondence relating to design. Basic criteria, data, or dealings with government agencies should be filed here.

Cost control files are not particularly voluminous but nevertheless are important in the control of the project. Cost reports, cash flow calculations, and other such back-up information are kept.

While all of this data is normally sorted in computers, hard copies should be filed separately in the project files.

Estimating files fall into the same category as cost control files. However, back-up data, such as quotations, are usually kept by the estimators for easy reference.

Construction files are kept at the site and are usually subdivided like the project files. The construction files kept in the project office are those relating to the interface between the project office and the site. At the close of the project, the site files are returned to the project office to form an overall record.

Drawing files are kept by each of the disciplines. Original drawings (transparencies) should be kept in a vault or in fireproof containers. Microfilms of the drawings are often kept in a separate office for security reasons.

Stick files are normally maintained. These are kept on a rack for working reference and can be marked up with colored pencil to indicate revisions that should be made. Coordination prints are circulated as required to all disciplines for comments and corrections. Changes resulting from this are transfered to the stick file and the coordination print is filed as back-up.

Record prints are kept of all drawings that are officially transmitted out of the office. A transmittal file is kept to log such movements.

Vendor drawings are filed separately.

In general, all filing can be done alphabetically, but misfiling is rampant when this is attempted. It is better to create a numbering system. Cross-referencing is extremely important. By storing file lists, reference numbers, subjects, and categories on a desktop computer, searching and retrieval can be simplified.

DISTRIBUTION LISTS

It is essential in organizing a project to prepare a distribution list, or series of lists, to control the day-to-day flow of paperwork. A master distribution list is a very cumbersome thing to draw up, and creating it always meets with resistance. Yet it forces management and the rest of the team to come to grips with who needs what from whom.

Noses get put out of joint when people sense that they are not being fully informed. Also, the invention of the photocopier has spoiled most of us. We expect copies of everything. If you take a democratic approach to preparing your distribution list, then you will find that everyone wants everything and that you can support an entire printing company by giving in to everyone's wishes. You can satisfy this dilemma by circulating copies of data rather than distributing copies. Also, take a hard line on a "need-to-know" basis.

Having made your lists of groups, what they do, and what they need, you can rough out a master distribution list. Do not make the mistake of thinking this will be an easy task that you can knock off in a day. Remember that it is the bureaucratic control of your project and how well it is done which will determine whether you have clear-cut communications or a blizzard of paper. Also, keep in mind that the distribution list is a living document that will evolve over the life of the project. Figure 15.1 is a partial master distribution list for a medium-sized project. Note that space has been allowed for additions.

Once you feel that your distribution list is in shape to use on the project, make an estimate of what it will cost you for paper. You will be shocked.

COMMUNICATIONS

The Telephone

We cannot survive without telephones. They have become so much a part of our daily lives that we hardly know we are using them. They have become an extension of ourselves. People slam them down in anger. They eat into them. They talk into them, look at them in disbelief, tuck them under their ears, leave them dangling, leave them off the hook, ignore them, walk around with them, drive with them, fly with them, use them for decorations, and rip them off the walls. Mr. Milquetoast becomes an aggressive salesman over the phone. We propose over the phone. We talk glibly with complete strangers thousands of miles away. We negotiate. We breath heavily. We change our personalities.

The telephone is the least expensive, most effective instrument of communication that we use. It is a wise thing to provide the best system available on your project and to encourage full use of it by your staff. Unfortunately, it is frequently abused. To use the telephone effectively, here are some suggestions:

- Do not skimp on the number of incoming lines. A busy signal loses money.

FIG. 15.1 — BEERBELLY BREWING COMPANY — MALTHOP PROJECT MASTER DISTRIBUTION LIST

Project No. 468

C = CIRCULATING COPY
O = ORIGINATOR
A = AS APPLICABLE

		MGT.			ENGINEERING						
		Project Manager	Eng. Manager	P M S Manager	Process	Mechanical	Civil	Architectural	Electrical	Automation	Municipal
MANAGEMENT	Project Instructions	O	1	1	C	C	C	C	C	C	C
	Incoming Corresp.	1									
	Letter to Board	O	A								
	Minutes of Project Mtgs.										
	Punch List	O	1	1							
	Time Sheets										
PROCUREMENT	Bidders Lists	1	OA								
	Request for Bids	1	O		1A	1A	1A	1A	1A	1A	1A
	Bids	1	1		1A	1A	1A	1A	1A	1A	1A
	Bid Analysis	1	1		OA	OA	OA	OA	OA	OA	OA
	Purchase Orders	1	1		1A	1A	1A	1A	1A	1A	1A
	Contracts	1	1		C	C	C	C	C	C	C
DRAWINGS	For Comment		1		OA	OA	OA	OA	OA	OA	OA
	For Approval		1		OA	OA	OA	OA	OA	OA	OA
	For Bid		1		OA	OA	OA	OA	OA	OA	OA
	For Construction	1	1		OA	OA	OA	OA	OA	OA	OA
	Approval Vendor Print	5			1A	1A	1A	1A	1A	1A	1A
	Final Vendor Prints	5			1A	1A	1A	1A	1A	1A	1A
	Stick File				1	1	1	1	1	1	1
	Vendor File Stick				1	1	1	1	1	1	1

FIG. 15.1 — CONT.

C = **CIRCULATING COPY**
O = **ORIGINATOR**
A = **AS APPLICABLE**

PROCUREMENT							ESTIMATING				COST CONTROL		
J. Snodgrass	Joe Piazore	Inspection	Expediting				Bill Pencil	Sam Guess	B.S. Louder	L. Sonofagun	J.L. Lost	T. Cashway	B. Check
1	C	C	C				1	C	C	C	1	C	C
1							1				1		
OA													
5A	C	C	C				1				C	1	
O													
OA	C						1				1	C	
O	C	2	2				1				1	C	C
1													
	O	2	2										
	O	2	2										

ACCOUNTING			P&S		CONSTRUCTION				P	ADMIN.		
J. Smart			L. Time	T.T. Short	I. Bild	O.T. Grante	Cal Lapse	F.L. Down	Payroll	Ian Dink	J. Drinkwater	S.S. Ship
C			1	C	1	C	C	C	C	C	C	C
					A							
1			1		1					1		
									✓			
					OA					C		
					OA					C		
										C		
					OA	1				C		
					1					C		
					O	1				C		
					2							
					5							
					12							
					12							
					12							
					1							
					1							

- Make a speaker phone available in a conference room for conference calls.
- Make sure that a telephone will rarely ring unanswered by providing access to a switchboard operator, secretary, answering service, or a paging service.
- Control long distance calls by using a system that bills calls to individual telephones.
- Encourage people to jot down a record of decisions or commitments made over the phone and keep these on file.

Telecommunications

Telex, TWX, and other telecommunications devices are, like the telephone, essential to our way of life and generally can be justified on a project. Use of the telex is often abused in the following ways:

- Messages are typed twice — once on a typewriter and again on the teletype. This is usually to account for bad handwriting. A little care in handwritten draft could avoid this, however, and only in the extreme cases, where handwriting is impossible to read, should the message be typed first. Encourage people with bad handwriting to take up medicine.
- Messages are often written in narrative form and end up reading like *War and Peace*. Encourage the use of point form.
- Telexes are sometimes used to get attention instead of to save time. Use the mails.

Telecopiers   *— and then came FAX !)) ! WM*

While they are convenient for transmitting graphic messages, telecopiers are usually not cost effective. But a construction manager waiting for sketches to arrive by courier would probably disagree.

Photocopy machines

We have already touched on the high cost of distribution. Photocopying is another function that we tend to take for granted. We copy everything: money, bills, letters, photographs, certificates, bonds, quotations, jokes, and club constitutions. Sooner or later some clown in the office will photocopy a hand or a face or some other portion of the anatomy. Control measures can be taken to limit illegitimate use of your equipment. Larger projects can assign an operator. There are counting devices and special keys that limit access to the photocopier. All of these methods may or may not be cost effective. The best protection against abuse is a busy office.

Typing

When oral communications don't work, we put it in writing. To give it effect, we have it typed. The printed word, it seems, has power.

Speedy memos are probably the most efficient transmittal method for the written word in offices today, yet we use them reluctantly. We have reams of material typed even if it is only for the record and not likely to get read. Encouraging the use of handwritten material, especially among friends, is an effective way to reduce costs. But more importantly, it saves time.

Typewriting machines have become increasingly more exotic and cost effective, especially where long drafts of legalese and technicalese are likely to require frequent revisions. The word processor is a blessing, yet it is in danger of being phased out and replaced by typewriters that plug into microcomputers or microcomputers that also serve as word processors. Before equipping your project with exotic equipment, you have to take into account the life of the project, and you must do a cost analysis in view of that life.

CHAPTER 16

MODUS OPERANDI

MEETINGS

To paraphrase a section from a previous publication by this author:

The size of project meetings increases as the square of the capital cost of the project and the cube of the number of financial partners.

Large meetings are characteristic of modern management. They are extremely inefficient, yet they seem to continue to increase in size and frequency. Expenses are astronomical if you consider the manhour costs and the practice of providing visual aids (the larger the meeting, the more exotic the aids) and adding luncheons or snacks. A typical megaproject meeting can have thirty to fifty participants. If you work out the hourly cost, based on, say, forty people earning an average of $40,000 per year, the cost becomes:

$$\frac{40 \times 40,000 \times 1.8 \,(\text{for burden and overheads})}{1,950 \text{ hours}} = \$1,476.92 \text{ per hour}$$

This doesn't include coffee and doughnuts!

One could argue that multimillion-dollar decisions are made at these meetings and that the cost of the meeting relative to the capital cost of the project is, therefore, insignificant. In most cases, though, large meetings tend to stratify into parochial debates. Few decisions are made unless these have been preordained and are simply opened for debate.

One executive walked out of a major project meeting in disgust.

"They flew all this high-powered talent in from throughout the country," he said, "yet they've spent the last two hours debating insignificant items like whether to ship the nails to the site in kegs or boxes!"

More often than not, the topic on the floor of large meetings is of concern to only a few of those present. The others who are obliged to sit through the discussions usually practice the fine art of yawning with the mouth closed, a skill that one requires up to a certain age, after which one merely falls asleep.

The psychology of holding large meetings is to give team members a sense of importance, of being party to major decisions, whether they actively participate in making the decisions or not. There is nothing worse than hearing of a decision that has been made that you were not a party to, particularly if you do not agree with the decision or feel that you should have been consulted. People find it insulting to be left out and they tend to lose interest in their jobs as more and more decisions are made without their involvement. It is for this reason more than any other that large meetings are held.

While politics are important, decision-making does become difficult if an objector decides to dig in his heels at a large meeting or if there are a variety of opinions on the same topic. The debates rage on at $1,476.92 per hour! A common technique used to cut off the argument is to form a committee from among the protagonists and charge them with sorting out their differences in a separate meeting to be held forthwith.

A better technique is not to use large meetings for decision-making at all, but to use them to present decisions that are already made. This will provide an opportunity for opinion to be voiced. Decisions can be made with smaller, more easily controlled groups and published in minutes which can be ready by interested parties. In fact, you can differentiate openly between the two. As a guide, meetings can be classified as follows:

information meetings — more than ten people attending
decision meetings — up to ten people attending, but the
fewer the better

Periodic meetings are essential to coordinate activities on a project. Otherwise the tendency is for interdisciplinary decision-making to bog down. It is as important to insist on regular meetings when things are frantic as it is when things are slow.

Naturally, when business is brisk, team members are reluctant to take the time to attend project meetings. If you do not insist on them taking the time, however, there is the risk that some decisions will be made that are difficult to reverse or that decisions that should be made are neglected. It is even more difficult to get people to attend meetings when times are slow. If you try to force people to attend under such conditions, be ready to be concise and to cut off the meeting in five minutes, if necessary, if nothing is happening.

One company religiously held a weekly meeting on a large project. There were about thirty in attendance each week. The technique used was to have each team member give an oral report to the project on his particular area of activity. Reports were prewritten and would be gathered for publication as minutes. Now, it would be very embarrassing for anyone there to not have anything to report, and, as a result, there was a lot of barnyard stuff slung about, particularly during periods of slow progress. Nevertheless, the meetings continued regularly. At one meeting a team member decided to liven things up and began his report with a joke. No one laughed, and he became quite embarrassed. He mentioned this to a fellow team member after the meeting, who asked, "Which joke?" When our man repeated the story, his friend broke up in laughter. So he decided to deliver absolute nonsense in his report the next week to see what reaction he got. There was none. If you force people into a regular routine, it is essential that it be an active, meaningful one. If the project manager in the previous example had been *conducting* a meeting instead of merely calling one, everyone would have laughed at a good joke. Instead, good manhours dribbled away to the tune of sawing logs.

At the other end of the spectrum there are individuals who regard meetings as a bitching forum, and there are others who have Shakespearean delusions and enjoy hearing their own voices. This is their opportunity to go on and on ad nauseum in a grand performance. Sometimes the firm hand of the chairman of the meeting can squelch the culprits, but there is a strange phenomenon that occurs in the midst of such gatherings that tends to augment, even support, oratory that has gone rampant. Often even the toughest chairmen fail miserably in their attempts to stop the flood of verbal diarrhea. If you want to limit a one-hour meeting to one hour in the face of these odds, hold it at 11:00 a.m. Grumbling stomachs will prevail.

Weekly meetings are an important and necessary form of communication that should not be underemphasized as a coordinating element in project management. In order to maximize the effectiveness of these meetings, it is imperative that communications throughout the project be sharp, so that surprises are minimized. One way to highlight important happenings is to maintain a weekly list of significant events that have taken place on the project and to include them as a section of the minutes. This list should be made up of one-sentence statements, e.g., "Decision taken by electrical group to locate substation in NE corner of the plant."

Punch lists should also be prepared and distributed and become a highlight of a weekly meeting. In fact, a punch list should be distributed regardless of whether or not a meeting is held.

THE PUNCH LIST

One of the most effective tools for controlling and maintaining momentum on a project is the punch list. Figure 16.1 is a typical example. Punch lists should be kept brief. Items that are critical to the progress of the project and that have priority over everything else should appear on the list. Anything else should appear in the general minutes of a weekly meeting.

Note that the item sequence numbering in the list carries the week number as a prefix. The reason for that is to emphasize delinquent items. Note that no due dates are given on punch list items. All are ASAP.

An astute project manager will place a great deal of stress on the importance of the punch list. He will use it to lean on responsible parties so that stubborn items will disappear quickly from the list. While discouraging the addition of frivolous items, he will encourage the players to use the list to expose teammates who are holding up progress. In figure 16.1, for example, the construction and electrical departments are in hot water for having stale

FIG. 16.1

PUNCH LIST   **Week No. 14**

NAUGHTY NIGHT HOTEL PROJECT

NO.	ITEM	BY
3-7	TAKE SOILS SAMPLES AT BRIDGE	CONST.
5-12	ORDER TRANSFORMERS	ELEC.
10-10	DECIDE ON AIR CONDITIONERS	MECH.
10-13	BID LIST ON ELEVATORS	MECH.
14-1	SIZE MAIN FURNACE BEAMS	CIVIL.
14-2	UPDATE SCHEDULE	PMS.
14-3	CHANGE WINDOW DESIGN	ARCH.
14-4	MODIFY MEZZANINE	ARCH.

items (items 3-7 and 5-12) on the list. The mechanical department is also in danger of being accused of holding up the project for items 10-10 and 10-13.

To utilize the punch list most fully, emphasize its urgency. If more than one office is involved, then the punch list should be issued by telex, and it should always be issued the same day that it is created.

One cannot overemphasize the need for tight control of this document. A sure sign that a project manager is losing his grip is when the punch list begins to grow in length and to contain a large number of stale items. Quite often, when this kind of pressure starts to build up, items are removed by partial completion. In item 5-12, for example, the electrical department may declare that the transformers are ordered when in fact, the bid analysis may have been done, the supplier selected and even informed of his success, but no order actually placed. Watch for this escape maneuver and insist on proof of completion.

All weekly meetings should begin with a review of the punch list and should end with an update issue of the list. In the event that there is no meeting, the project manager or his delegate should review each item with its responsible player *without fail*.

TRAVELLING

A lot of travelling can be avoided by the use of conference calls. Visual conference aids such as television are being used more frequently than ever. Perhaps someday holograph images will be sent electronically so that you can have a meeting with someone else's image and review a set of drawings that are real at one end of the world but only an image at the other. The ultimate future solution may be teletransportation of one's body by breaking it down into atomic particles and reassembling it at some other location! This conjurs up a whole gamut of things that could go wrong like winding up in a bordello instead of a board meeting or having your head located where your foot should be and vice versa.

In the meantime, we are stuck with travelling in our work. Travelling is considered a nuisance by some people and an escape from routine by others. In nearly all cases, it is an excuse to go first class whenever possible, the rationale being that when one is on a trip, one should live as well or better than one does at home. And besides, you are only being paid for an eight-hour day! Most project managers find that rationale hard to combat. Travel costs have a tendency to escalate with time elapsed on the project. It is not that the basic costs escalate but that the rationale does! If hard and fast rules are not laid down at the beginning of a project, the budget will fall victim to this phenomenon. And so will the project manager, who will be plagued by all sorts of expense claims he never dreamed of — special laundry, lost personal effects, emergency clothing requirements — the list is endless. If your accounting or tax people will allow it, per diem living expense allowances are always easier to handle than itemized accounts, and the hassle with employees then is reduced greatly. If your traveller wants to eat wieners and beans for a week in order to have champagne and caviar for a day, let him do it.

Justification for travel is always a high profile item. So is justification for not travelling. One manager proudly announced that he had little need to travel. All he had to do was punch a command into a computer terminal, and he got all the information from the site that was needed. Saved him three days of travel. Yep. Trouble was, he missed seeing the tongue-in-cheek of the guy who was feeding the data into the computer at the other end, and didn't find that out until he had blown his budget. There is something about personal, eyeball-to-eyeball contact that makes travel worthwhile, even for brief meetings.

A junket can be defined as a frivolous trip, and not all projects are able to escape the odd one. The following are a pair of examples.

A major iron ore project had been experiencing a delay in settling the design concept for distributing iron ore concentrate to several surge bins. Traditional methods, such as trippers and shuttle conveyors, were proving costly in building steel, and so the project manager directed the design team to use belt plows, which would be quite inexpensive and would reduce the size of the building by a considerable margin. The mechanical group leader, however, voiced strong objections, insisting that the moisture content would be such that the concentrate would stick to the belt conveyor "like fresh cow dung." He gained support from other key members of the team.

The project manager faced a dilemma. If he went with traditional design methods he would almost certainly exceed his budget. If he authorized further design study, he would lose his schedule and the results of the study might even have them revert to traditional methods. Then he would lose both his budget and his schedule positions. After a half day of frantic telephone calls made to try to find an operating example, the project manager came up with a plant in Missouri that was using belt plows in a similar application. Still, the project team was not convinced. In a final desperate move, the project manager sent the mechanical group leader and two other members to see the belt plows.

Since it was the height of the holiday season, the trio had to travel first class in order to get reservations. A half day later they staggered off the plane in St. Louis, having done their best to deplete the entire stock of the first-class cabin's refreshments. Now, the hotel that had been selected had a gourmet kitchen and great entertainment. The trio got to bed at two in the morning, almost forgetting that they had to be up at five for the three-hour drive to the plant. Their visit had become an ordeal, and despite having to agree that the plows would work, they faced a dressing down from the project manger when they returned for: 1) arriving at the plant late, and 2) exceeding their expense account limits.

When they finally returned home, they went to the project manager to face the music. Maintaining a stern façade, the project manager could hardly contain his delight. The $3,000 trip had saved $500,000 and his schedule.

Sitting in his office in San Diego, an owner's manager and his consultant were planning a trip to Italy to negotiate with the supplier of process equipment for his new plant, which was in the early stages of design.

"I hear," said the owner, "that there's a plant being built in the Sultanate of Oman."

"You're right. In fact, it's almost finished," said the consultant. "I believe that they're commissioning the equipment about now."

"Can you arrange a visit? I know it'll just be a junket, but after all, Oman isn't too far from Rome."

"Of course. I'll make the arrangements."

A couple of days later, the consultant returned to inform the owner that permission to visit the plant at Oman had been requested and that a positive reply was expected soon. When he arrived at the owner's office, he found him sitting with his feet up on his desk, staring at a map of the world that hung on his wall.

"Shoot!" he said. "Oman is halfway around the world. Wouldn't it be a hoot to come back the other way? Heck, I've never seen the East."

"Can you authorize it?"

" I'll give it some thought."

Despite the risk of appearing careless of company funds, the pair made the trek to Italy and then to Muscat, Oman, where they toured the new plant in 120 degree temperature. Then they proceeded to enjoy the delights of the Orient, enroute to San Diego. Half the trip had turned into a junket. Two days after their return, the owner and consultant were putting together their expense accounts.

"That trip, great as it was," growled the owner, "cost over $10,000!"

"I've been giving a lot of thought to that," said the consultant. "It was miserable climbing all those stairs in 120 degree heat in Oman. Your plant, as presently envisaged, is nearly identical to that one, and the climate can be nearly as bad. What kind of maintenance can you expect on equipment at the upper levels of the plant?"

"There's an elevator."

"If it's working or if it's not already in use."

"What are you saying?"

"It strikes me that we've approached the plant design all wrong. You've got plenty of real estate. Why don't we spread the design out, make a low profile design? Come to think of it, what with the high earthquake factor in California, it should be much cheaper."

"But that would be a departure from the traditional design," said the owner.

"Traditional does not necessarily mean best."

After a new conceptual layout and estimate had been made, the consultant beamed at the expected savings of nearly $2,000,000.

"That junket," said the owner, "cost $10,000, but it saved two mil. What can I say?"

"Even junkets can pay off," smiled the consultant, remembering his delightful nights in the Orient.

DESIGN AUDITS AND PANEL REVIEWS

Earlier, in Chapter Two, we spoke of using design audits and panel reviews to monitor technical quality and to ensure that the project was headed in the right direction. In the same chapter, we warned about putting the wrong people in place to make judgements about the staff who are doing the job. As one engineer put it, "I have reservations about someone who thinks he can know more about a job in a few days than I do after six months!" If you do not bring in respected mentors, you will get just that reaction.

Design audits can be of two types — intensive or judgemental. An intensive audit is the more severe type and is used where the public is in peril if things go wrong, such as in the nuclear industry, chemical plants, or explosives plants. In these cases, everything in the design is checked by an independent group, starting with basic calculations and continuing through to manufacturers' certificates. Judgemental audits, on the other hand, are done by engineers who have had direct experience over the years. They will study layouts and drawings and can tell by experience whether or not things look right. When they sense an error or an instance of questionable judgement, they will probe deeper, asking to review calculations and criteria. Quite often a judgemental audit will uncover things that an intensive audit will not. In an intensive audit, the relationships between parts of a design often are overlooked in the quest to uncover errors in detail.

Panel reviews usually cover overall aspects of the project and are used to check engineering, design, procurement, estimating, cost control, and construction. In other words, they evaluate the general health of the project. Sometimes panel reviews become a case of who is kidding whom. A skillful project team will do a presentation that is as dazzling as a Las Vegas review, really a sales scam on the panel. The panel, on the other hand, sometimes are given a mandate that they really can't handle, mostly due to other pressures, and so they cannot delve into the details that they need to make a sound judgement. What normally happens in that case is that everyone carries on a kind of charade, after which they all stand around drinking martinis as the band plays on. With a little luck the project will survive on its own.

PROFESSIONALISM

As engineering and construction have developed over the years, so has the threat of litigation. The growth of that threat can be measured by the increase in the volume of paperwork, the trend toward decisions made by committee, and the language of contracts, which tends to lay the blame for anything that goes wrong on anyone but the engineer and the construction superintendent. Organizations, too, have tended to become more matrixed. Generally, corporations accept responsibility for their engineers. However, the individual often feels subject to corporate pressures, and perhaps those, too, tend to foster a pass-the-buck attitude.

If management of capital projects expects to improve in the future, then it must instill a greater sense of individual responsibility in the members of the team. As a professional, an engineer should be no less responsible for the work he does than a surgeon who makes the final cut. He may be assisted by anaesthetists, nurses, technicians, and interns, and he may be supported by a fine hospital administration, but it is his hand on the knife. He is being paid well to see that it is steady. So it should be with engineers.

To maximize efficiency, we must maximize individual responsibility.

ARTIFICIAL EMERGENCIES

Keeping momentum up in projects is a universal problem. Probably the best, most efficient use of personnel comes in the beginning stages, when enthusiasm runs high. This enthusiasm can be sustained by regular meetings.

According to what seems to be a general pattern, enthusiasm

wanes when about two-thirds of the project is completed. The project manager may find himself lacking in spirit, and you can bet that he will have plenty of company. When that malaise sets in, along with it comes a creeping deterioration of the budget and schedule. If monitoring of these is maintained, such that a deviation in the S curve is noted, then there is usually a flurry of activity as the panic sets in. This flurry is often sufficient to bring the project back on curve, provided that the fear of calamity has taken root in members of the team. The problem can be, however, that in getting off the curve, the project is vulnerable to failure if the slightest upset is experienced.

Astute project managers will expect and therefore watch for this slump in project momentum. One of the most common strategies when enthusiasm wanes is to create an artificial emergency. A legitimate and useful exercise is to call for an estimate update. This will require, of necessity, a schedule update and will bring into the act all of the characters of the play. If the update is reinforced by a slightly unrealistic due date, the attention of the group is even more focused.

Using milestone contracts such as mechanical installation, for example, and creating a reason for early preparation can spur the whole procurement chain onward, thus generating enthusiasm.

However, in using such artificial emergencies, a project manager should realize that he can get away with this strategy only a few times throughout the project, after which his "emergencies" will go unheeded.

GAMES PLAYED IN THE INDUSTRY

When you are planning to build a plant or a refinery or to open a mine, there is an immediate faceoff between those who wish to make money with the project and those who have to run it. The operators and maintenance people want to spend all that is necessary to make their own lives easier. The investors want to spend as little as necessary in order to reap a maximum return. Usually the investors ask "How much?" The operators quote, and the game is on.

Typically, each side is revved up in a different way. The investors, operating on skimpy information, usually have been turned on by ROI figures that will instantly transform them all into North

American sheiks. The operators, caught up in this fever, all want to drive Cadillacs after they have made the operation work. Reality sets in down the road when hard costs or markets are estimated. Investors who have been flamboyant in their approach now become concerned. Operators who have been cavalier in their design requirements now begin to worry about how difficult the plant may become to operate.

About this time, the project manager gets into the picture. He has to strike a balance somewhere. If he designs and builds a cheap plant, the investors may be ecstatic, but the operators will come down on him like a herd of elephants. If he is wise, he will realize that if an operator doesn't want a plant to work, he will damned well see that it doesn't. And if it doesn't, guess who gets the blame? The operator? Hell, no. He'll just say that the plant was not designed properly and has not been built correctly. And guess who was responsible for *that*? Clearly, a wise project manager had better get the operators on his side.

As far as the investors are concerned, the project manager can usually expect them to reel in horror when he tells them how much the plant will cost, no matter what the figure. If he is a typical manager, though not necessarily a wise one, he will try to stick the investors with the highest cost and longest schedule he can get away with during the budgeting stage. He runs the risk, however, of scaring the investors away from the project.

While the project manager is walking the tightrope between investor and operator, he must be aware that his reputation is on the line. If he sets up a fat budget and undercuts it by a wide margin, he is no further ahead in the credibility department than if he overruns. He has done no one a favor because he has simply tied up funds that could be used elsewhere. Fortunately, if he is able to push through a bulging budget, he is usually saved from the embarrassment by a kind of Parkinson's Law of finance that states:

> *Money will be spent to fill all the voids in a budget — and then some.*

Obviously, the true-blue project manager is the one who can accurately predict the cost and schedule of a project, but there is no doubt that making the last bits of the puzzle fit is a lot easier if you're dealing from a fat deck.

Astute investors usually recognize this budget-fattening syndrome and coerce the project manager into a lean budget that is sometimes leaner than makes good sense. The project manager then has little chance to be a hero by underrunning the budget, except by further decreasing the quality or scope of work at the risk of losing the faith and cooperation of the operating team. A very ticklish situation indeed.

What normally happens in such a case is that the project manager finds himself overrunning his budget early in the game. He can go back to the board, pleading an underestimate. There is an unwritten rule in business, however, that you can get away with that once, but you had better not try it a second time!

One project manager, seeing no other way out, decided to delete all the stand-by equipment from the plant, much to the chagrin of the operating people. To his surprise, the operating team stopped fighting him in a very short time and displayed a cooperative attitude that was puzzling. During the final construction phase, just prior to commissioning, the project manager made a visit to the site and was shocked to see all the stand-by equipment sitting in the warehouse! On investigation, he discovered that the plant manager had bought the equipment out of his operating budget. Naturally, the operating budget had been overrun substantially. But it was a smaller budget and attracted less attention from the board. And when the board did complain, guess who got blamed?

Getting money out of a board of directors or investors for a simple budget overrun can be embarrassing for a project manager, particularly if he had a hand in the original estimate. If he continues to ask for money, it also becomes embarrassing for the board, who approved the budget in good faith. It reflects on their judgement. Traditionally, both the board and the project manager have gotten themselves off the hook by creating changes in the scope of work. A manager thus cornered will encourage the operating team to make changes and will inflate the costs slightly to cover losses elsewhere. It becomes much easier to approach the board for more money if the request comes with a list of changes that presumably will benefit the operation. By manipulating a deletion here and an addition there, a clever project manager can pull himself out of the soup.

Naturally, if the original project was conceived to cost, say, $100 million and it comes in at $150 million, there will be a lot of eyebrows raised, perhaps by irate shareholders. One of the oldest tricks in the book is to create budget updates or entire revised budgets during the course of the project, usually based on scope changes. The closer you can get the creation of the "new" budget to the end of the project, the better your chances of meeting it. Then you can make an announcement that "the plant was budgeted at $150 million and came in at $149 million for a savings in capital cost of $1 million." The project manager is a hero, and the board looks good to the shareholders. The odd shareholder who complains that the plant was supposed to cost $100 million will be told that *that* budget was for a different design entirely!

One may smile at all of this, but these methods are as old as prostitution and will continue, in spite of what this book or any other says. In the interest of improving project management in general, however, a project will stand a better chance of success if the project leader is insistent on a clear and frozen scope of work beforehand. He must seek this commitment from both operators and the board prior to spending hard costs. His goals then should be:

- to get a commitment for a clear, definitive scope of work;
- to stay on budget;
- to stay on schedule;
- to monitor progress to ensure the above.

If all else fails, well . . . play a few games.

CHAPTER
17

WORKING WITH THE CONSULTANT

SECRECY AGREEMENTS

One time, while working in an Eastern European country, a Canadian engineer was warned time and again about taking photographs of a proposed plant site which he was examining for a feasibility study. Several times during his tour, threats were made to confiscate film and/or his camera. Even so, he managed to return with five full rolls of exposed Ektachrome 64 containing photographs of the site of the proposed plant, several dozen tourist attractions, and a rocket installation. While having dinner with his client, with whom he had become friends and who was visiting his company's office in Montreal, he casually mentioned this fact. A look of horror came over his client's face, but after he had calmed down, the engineer produced an aerial photograph of the plant site that he had purchased from an agency in the United States for twenty-five dollars.

The point of this story is that in today's age of electronics, computers, and other sophisticated technology, there are very few secrets. Asking a consultant and his employees to sign a secrecy agreement, therefore, is very naive indeed. Yet the practice persists, particularly in large organizations.

If your aim in having a secrecy agreement signed is to *retard* the availability of information to competition, then you may be on the right track. If it is to *stop* the flow of data, however, then the most sensible advice is: Don't bother.

SCOPE DEFINITION/GETTING STARTED

You may have elected to use a consultant throughout the evolution of the project from feasibility through completion. For this exercise, it is assumed that you have completed your feasibility, have raised the financing, and are about to begin, after having contracted a consultant.

It is customary — and necessary — at this point to hold a kick-off meeting. This meeting should be conducted as an information meeting with all supervisory personnel of both the consulting group and your company. The purpose of the meeting should be to review the following, which should be included on a pre-published agenda:

- Identification of participating personnel
- Organization
- Scope of work
- Milestone schedule
- Short-term target schedule
- Budget
- Early commitment requirements
- Reporting expectations
- Manpower distribution and build-up

It goes without saying that the most important function of the meeting is to give everyone a clear picture of the project and which paths are to be taken to realize its success. Too often part of the group gets off to a flying start, while everyone else wallows around, not knowing what to do. All members of the team should know exactly where they are going.

Identification of participating personnel and organization of the project is an important first step in your kick-off meeting, but simply introducing people is not good enough. People forget names, titles, and functions, so it is a good idea to tie in the agenda with the personalities involved. This reinforces the individual's responsibility to himself and to those present as well as the relationships between the various groups.

It is essential in your kick-off meeting to review and reinforce the scope of work as per the contract. More importantly, you should emphasize to the project group as a whole that one of the primary objectives during the preliminary design phase is to firm up and further identify the scope of work. After all, the budget under which you are operating in the beginning is likely to be based on an order-of-magnitude estimate from a limited amount of information. Some flexibility can be exercised during this stage, therefore, to adjust the design according to cost information that has been developed, before hard field costs are committed. This is not license, however, to permit wholesale changes. In fact, it should be made very clear that changes of scope must be identified before they happen — as trend forecasts — and that actual work on changes cannot be performed without formal approval.

While the milestone schedule should be presented, a short-term target schedule covering a three-month period should be laid down, committing those present to definite, achievable goals. This is a chance to have the participants vocalize their commitment, thus reinforcing their own identification with the organization.

An overall review of the budget must be made. Often this overview is not regularly presented to the whole group. However, if you expect people to commit themselves to it, then they should know, at least in general, what is in it.

Early commitments are sometimes required and should be identified if they are expected to affect the critical path. For example, the transformers may have a delivery of forty-two months, and your overall construction schedule may be forty-eight months. A purchase commitment may have to be made out of sequence in this case, estimating the size of the equipment before all the data are available.

Procedures for reports should be made clear. There is no point in receiving a monthly report if you require a weekly report or only a quarterly report. Even though all of the participants in the meeting are not obliged to report to you, they should know who is and to what extent they will be involved themselves.

Discussing manpower build-up can be an effective way to control and emphasize the fact that too rapid a build-up will result in idle time, and too slow a build-up will delay the schedule. Although it is important to sensitize the group to this fact, it is just as important to realize the impact a too rapid or too slow build-up can have on you and your own team. If you are slow in decision-making, you could end up wasting expensive manhours. A good consultant will be quick to point that out to you; a bad one will laugh all the way to the bank.

Scope Changes

Even on fixed-price contracts, consultants are sometimes slow to obtain approval for scope changes, and many consultants, despite having a high level of competence, do not use or understand trend forecasting. If you sense that this is the case, then you had better take action, otherwise you may find that the project has drifted merrily along towards an overrun. The project should be monitored for trends and scope changes on a weekly basis. Scope change approvals should be made by you monthly. If changes are not submitted to you, ask for them, or you will receive a bundle of them when you least expect it, and worse, you will feel committed to approving them. If your consultant does not have a trend-forecasting system in place, you could elect to put in your own man, bearing in mind that he must have a mandate to report on trends generated by your own team as well as the consultant's.

Scope changes are of two sorts: changes in the consultant's contract and changes in the project itself. Changes in the project do not necessarily affect the consulting contract, although they usually do. As a result, it is normally best to report both in the same document.

Insist on having a running account made on the budget *at the same time* that change orders are presented for approval. If the consultant's change order documentation doesn't include budget information, then have it changed. It may appear easy to approve a minor increase in cost, but when it is put next to a running account, it may make you pale.

And lastly, when a change order involving a cost increase is presented for approval, always ask the question, "Where else in the project can we make a compensating change to offset this increase in cost?" If you look hard enough, you may be surprised to find many tradeoffs.

COMMUNICATION/LISTENING/ CONTRIBUTING

Because of the fact that the scope of work in the beginning of a project is flexible and your goals still have to be defined to the point where the design and scope are frozen, the consultant is faced with the task of learning more about exactly what you want. Sometimes even you don't know what you want, and he has to try to evaluate your requirements and match them up with reality. You and your consultant have a truly synergetic relationship: you have to work together.

Common practice in industry is to have at least one client representative with the consultant on a daily basis, occupying a permanent desk in his office for the duration of the project. Some consultants do not like to do this, since there is the possibility of day-to-day interference with progress. Also, there may be objections from other clients who are sensitive to other companies having visual access to their work. These objections can be overcome by isolating your project to a specific area of the consultant's office. Alternatively, the consultant can work from your premises or a separate facility can be leased.

This should not imply that you have a right to complete freedom in the consultant's office. On the contrary, they have internal matters to concern themselves with and you should respect their privacy. Constant contact, however, is highly desirable.

As soon as you occupy part of a consultant's office, it is quite important to establish exactly who is in charge. Certainly, you are responsible if the project goes sour because the consultant thought you were in charge or vice versa. In your kick-off meeting, you should emphasize that the day-to-day running of the project is very definitely the consultant's responsibility and that you are there to provide information and approval, unless, of course, you feel obliged to take control because of the consultant's incompetence.

There is a feeling in many industries that consultants do not know the industries they serve well enough and many consultants think that clients are their own worst enemies when it comes to running projects. A lot of this feeling is caused by lack of communication or breakdown of the same. Both parties must realize that they are better at what they do than the other. Operators know operations better than consultants; consultants know projects better than operators; and neither is perfect at what he does. It is the coming together of these skills — the synergy — that is going to give good results. In fact, when you think about it, by definition a successful design and project execution is a total compromise. The process people want ideal flow conditions, which are compromised for layout, which is compromised for ease of operation, which is compromised for access for maintenance, which is compromised for cost, and so on and so on. By orchestrating all of these demands, an optimum solution will evolve.

Providing an atmosphere for effective communication is largely your job. If a consultant has a budget, schedule, and basic scope of work, he would be most happy to carry out your project without any interference on your part whatsoever. Remember, though, that the results you get will be entirely the consultant's interpretation of your needs. If you constantly contribute to the consultant's job, by having regular meetings with your operating and maintenance personnel, for example, then you can effectively influence the project. There is a tendency for owners to forget, however, that the consultant can bring to your project operating and maintenance experience from other clients; that, although the consultant's expertise is primarily in project work, he is exposed to problems similar to yours frequently. Closing the other half of the synergetic circle, therefore, is best done by effective listening.

When the Consultant Gets Sold Down the River

Clients who do not listen and who impose their wills on a consultant often put him in a very awkward position. If a consultant does not believe in what you are insisting he do, then a very sour relationship can develop. It is difficult to say no to someone who is paying the bills.

One young project manager had been assigned the job of directing the conceptual design of a shiploading facility. "Pay special attention to the railyards," said the client's project manager. "We must have capacity for storing at least 180 railcars."

After studying the land availability, shed layouts, docks, and shiploader arrangements, the consultant declared, "It isn't practicable to store such a vast number of hopper cars. We suggest that you look, instead, at faster turnaround so that you'll need only a surge area of perhaps fifty railcars."

The client's project manager, a stubborn individual, suggested that the consultant knew little about shiploading operations and that he had best confine his efforts to fulfilling his client's wishes and not to trying to reinvent the wheel. After all, this was merely the initial phase of a very lucrative contract for the consultant, and he should not risk offending the client. A consultant could be replaced!

Three weeks after the conceptual design had been completed, the consultant's project manager got called before the president of the consulting group.

"We've been removed from the project," he said testily. "What have you got to say?"

"I don't understand," said the project manager. "Why?"

"Their vice president called on me this morning. He said that they showed your design to the railway people. He said they laughed. You have far too much trackage in your design, and the amount of shunting required is unreal."

"But," said the project manager, "their project manager insisted. He told me they had to have storage for 180 cars!"

"Nonsense," said the president. "It's the railway's responsibility to store the cars. All the shiploading company needs to provide is a surge of twenty or thirty cars at a time."

"I *told* him that!"

"Then you should have stuck to it. It's better to be removed from a project for being right than for being wrong. We may have lost a contract over it, but if the yards had actually gotten built, our name would have been on it forever!"

It was too late to recover the contract because a new consultant had been hired. However, the project manager had a rule firmly imbedded in his mind:

Never let a client sell you down the river.

And as an owner, you have to realize that a consultant has an obligation to do a job in what he believes is the best way.

YOUR DIRECTIVES CAN BE A LICENSE TO PRINT MONEY

We talked earlier about who should be in charge of the project and suggested that, basically, it was the consultant's role to be in command, except, of course, when it is necessary to take the helm because you feel that the consultants are not doing a proper job. The risk involved is that you, too, may not be up to the job. Then what?

Many a consultant will feel miffed if control is taken away from him and will revert to "yes, sir, no, sir, three bags full, sir." He may take to making copious notes of your every decision and to following up each decision with a change order. Before you know it, the list of change orders will be endless. The same thing can happen if your project manager is too assertive at the beginning of the job. The consultant's project manager feels threatened by your directives and worries about being sold down the river, not just in design, but in budget and schedule as well. He will adopt a "protect-your-ass" attitude, which is okay as long as the project goes well.

In summary, it pays to establish a firm pecking order in the beginning, give the consultant his head, and monitor what is happening.

CHAPTER
18

COMMISSIONING AND START UP

WHAT IS COMMISSIONING?

Essentially, commissioning is the last stage of preparation for plant start up, beginning sometimes before installation and ending with a declaration that the plant is ready to go. Planning should begin well in advance of the installation of the first piece of equipment. People who should be involved are your operating and maintenance staff, your consultant's commissioning delegates or team, and suppliers' representatives.

During the procurement stages of the project, suppliers of operating equipment should have provided installation, operating, and maintenance instruction manuals. These should be catalogued and a composite operating and maintenance manual written for the entire plant. Interlocking sequences, control loops, piping circuits, and lubrication schedules should be highlighted.

Although the theoretical responsibility for correct electric motor hook-up rests with the contractor, a delegate from the commissioning team should at least witness determination of correct motor rotation. On projects that have a large number of electric motors, it is usual to buy these in bulk directly from the manufacturer. When this is done, it follows that installation of the motor is done by one trade and hook-up by another. When the hook-up is done, the coupling must be broken, the correct motor rotation confirmed, and the motor realigned and mechanically connected. One may argue that an installation acceptance certificate from the general contractor absolves the commissioning group of responsibility, but what is there to lose by personal contact?

If you have had the foresight to specify lubricants by brand-name or grade in your specifications, then the commissioning team will have an easier time preparing a lubrication schedule. Otherwise, they will have to match brands from the various operating and maintenance manuals and obtain the blessing of the manufacturers or risk voiding warranties. Once the schedule has been established, then ordering of quantities sufficient for initial filling and at least a year's operation should be done. Then, of course, actual filling with lubricants must be carried out.

Other consumables, such as coolants, liners, grinding media, reagents, and sealants should be treated in the same manner as lubricants. And, of course, the commissioning team should make sure that sufficient spare parts for the first year of operation have been ordered and that adequate warehousing is available.

Specialized process equipment may require the services of technicians or engineers from the factory to commission their designs. These people may also be required for installation and start up, depending on the complexity and/or sensitivity of the design. In these cases, formal written approval of installation should be obtained. Usually, an isolated no-load test is sufficient to satisfy the factory representative, but in some cases load tests are necessary to detect leaks, temperatures, and expansions prior to actual start up.

Other process equipment that is standard — pumps, conveyors, vibrating screens, valves — needs to be checked out by your commissioning team one at a time. Many companies will tag equipment "Ready for Start Up" to ensure that work is not duplicated.

Piping and ducting are normally flushed and cleaned by the installer to remove debris, cuttings and dirt. During commissioning, piping should be filled with compatible liquids or gases and pressurized to check for leaks. Special procedures for toxic and hazardous materials must be followed in accordance with published codes.

Electrical circuits and interlocking sequences should be checked using standard testing equipment. Control loops can be tested in conjunction with no-load tests of equipment.

Particular attention must be paid to safety devices under simulated emergency conditions, such as conveyor belt misalignment. Checking of systems for fail-safe features should be done, even though design criteria may have specified these. One example of a mishap took place when a microswitch on the cable drum of a shiploading crane got ripped off by a wild cable. The cable was supposed to simply trip the switch. However, the switch disappeared before it had a chance to do anything. As a result, the boom fell on some workers, crushed them, and was itself destroyed.

Calibrating of weighing and measuring devices is a time-consuming, meticulous job. Many devices may require at least partial loading or simulated loading. Belt scales, for example, use a calibrating chain. Other weighing devices can use static loads or hydraulic rams to calibrate them. In some cases, volumetric measurement is required to simulate operating conditions.

A plant should never be declared ready for start up until fire systems are charged, checked out, and signed off: in other words, commissioned. Similarly, safety equipment — eyewash stations, for example — should be in operating condition.

And finally, maintenance records should have been set up, ready for the first entry as the wheels begin to turn. In them, notes on deficiencies encountered during commissioning and corrective action that has been taken should be made in order to have a record in case of warranty claims.

Once all of these items have been commissioned, then a formal sign-off declaring the plant ready for start up should be made by the leader of the team and the project manager.

When to Bring in the Start-up Team

Too often we hear of start-up teams who come on the job ill prepared and late. This is as bad as a start-up crew being presented with a plant that is full of temporary wiring and jumpers because the project team wanted to meet a start-up schedule. The start-up team should begin work with the commissioning team and, in fact, can be made up of some of the same staff. Often a consultant can provide people who specialize in starting new plants, who have participated in many start ups, and who are aware of the pitfalls that can be encountered.

The Use of Models in Training

If your plant is in any way complex, you will appreciate having a model to acquaint start-up, operating, and maintenance personnel with the plant and to train them in its operating principles.

Usually a model begins its life during the design of the plant. It is then shipped to site for training, and finally, it is placed in a display case somewhere in the corporation. The model should be designed for easy breakdown, crating, and shipping.

If you feel that your plant will have a moderate-to-heavy turnover of personnel, such as is sometimes encountered in remote locations, you may be wise to keep the model on site as a permanent training device. If you do this, you should be prepared to keep the model up to date by having it modified to reflect changes that you may make to the plant from time to time.

CRUNCH TIME — START UP

Okay. The plant has been declared and signed off ready for start up. Heart palpitations can begin!

An engineer was inspecting a successful pelletizing plant in Michigan, hoping to learn about some new wrinkles in plant design. He spent a full day poking into every nook and cranny. The owners were very accommodating indeed. During his six-to-eight-hour stay, he saw only a single employee other than personnel in the control room and shops in that 120,000 square foot plant! And that man was merely passing through! The plant was completely automated. That may also be the case in your plant, but never, never during start up. In fact, your start-up crew should have quite a bit of depth in both talent and numbers.

Certainly, it is a rare process plant that starts up at the push of a button, comes up to full production, and continues ad infinitum. If your plant is modular or has distinct subplants, then your job will be much easier, because you can focus on one unit of the plant at a time. If, for example, you have a crushing plant and a shredding plant, each feeding surge bins ahead of a mixing and blending plant that in turn fills feed bins for further processing, then you should start each of these in sequence, prove that it can operate, and then go on to the next. It is not impossible to do all at once. However, you are spreading yourself over a larger area, and that means that more things can go wrong.

Personnel should be briefed on start up, of course, but also on shut down. Even if continuous operation is planned, the start-up proof of the plant must include the shut-down sequence. It is a good plan to operate for, say, six hours, then to shut down for debriefing. This may not be practical for high temperature operations. If it is not, then plan a one-week operation.

Position a responsible individual at a shut-down pushbutton station at each major piece of process equipment. If you have not had the foresight to install a shut-down pushbutton ahead of time, pooh on you: Get a temporary one installed, and put in a permanent one as soon as possible. Have fire and safety personnel on stand-by during the entire operation. If you cannot communicate directly to each area by walkie-talkie or telephone, then you will have to rely on audible or visible signals. There should be some system of communication from each work station to the control room to cover the possibility of a manual shut down occurring and the interlocking not working.

Once a section of the plant has been operating for a reasonable period of time, stand-by personnel can be shed on a scheduled basis. Many plants take weeks, even months, to come up to full production. Sometimes this is deliberate, in order to break in machinery and personnel, and at other times it simply takes that long to optimize the process. Others do not anticipate sales of full production for some time and take advantage of the interim to optimize operations and maintenance.

When Things Go Wrong

Even using off-the-shelf designs is not a guarantee that you will not have some surprises during start up. Despite the best brains, the best manufacturers, the best designers, the best builders, Murphy's Law will operate. Something, no matter how small, will go wrong.

An engineer once designed a lump breaker especially for smashing frozen iron ore concentrate that was to be stored on the ground. It had been hoped that an off-the-shelf design would be available in the industry, but there was nothing at reasonable cost. So the intrepid designer set out to create one. The client was not convinced that it would work. The designer's youthful zeal, however, won him over and the design was built and installed. At start up, its total running time was approximately ten seconds.

You see, the operator of the front-end loader dug beneath the concentrate pile in his first scoopful and loaded the precious design with granite boulders!

W.B. Henderson is fond of telling the story of the time a plant that he had managed finally reached full production. He was standing on a walkway with his client, both of them leaning on a handrail, staring peacefully at the smooth-running activity below, at one in their awe of North American technology. Just as the client was congratulating him, out of the corner of his eye he noticed a bulging polyethylene launder beginning to collapse and was horrified to see 16,000 gallons of muck flood the pumps below!

On a sadder note, there is the story of the first long-distance water pipeline designed in the early 1900s for installation in Australia and covering a distance of over 100 miles. After the pumps had been running for twenty-four hours, water had yet to appear at the discharge. The designer, beside himself with humiliation, took his own life. Six hours later, water began to flow from the end of the line.

It is to be hoped that your plant will not encounter too many problems. If it does, however, do not take the rash action mentioned above. With some exceptions, there is hardly a plant in the world that cannot be made to run, given enough time and resources.

Maintenance During Start Up

When start up takes longer than anticipated, and there is a frantic cry from the investment forces to get it going and up to production; when the floor is a mess because of a burst pipeline, and there are jumper wires hanging all over the place while the electricians try to sort out what's wrong with the interlocks; when the whole crew is on extended overtime, the most likely thing to suffer is general maintenance.

A little voice inside you says that once you get rid of this mess, you'll get right on it, and if not then, it will be caught up during scheduled shut down for maintenance six months down the line. And what the heck, the equipment is new, isn't it? If you listen to that little voice, brother, you are dead. You will never catch up with the maintenance.

New equipment, be it airplanes or apple-squeezers, needs maintenance from the day it gets out of the factory. In operating a plant, you cannot afford to get behind the maintenance curve — ever — or you will be behind it for the life of the plant.

When you plan a start up, plan for the worst. Set up your maintenance program in such a way that it is not vulnerable to being pirated for extraordinary start-up problems, even if that means hiring a contract maintenance crew to sustain you during the critical start-up period!

PLANT OPERATIONS

Theoretically, the project manager's job ends with a plant that has been started up and proven as an operating entity. The manager is remiss, however, if he has not devoted some thought to operations long before commissioning and start up. If, for example, he sees that the future plant management are not properly preparing themselves for the job of operating and maintaining the plant, then he should take steps to see that they are prepared, despite the political ramifications of that move. Because if the plant runs into serious operating problems, guess who will get the blame?

CHAPTER 19

LOOKING BACK

AS-BUILT DRAWINGS

A disgruntled structural designer with a consulting company once circulated a perspective sketch that he had made while waiting for a particularly indecisive client to make up his mind about where he wanted an access stairway built. His sketch showed a building column with a section of it boxed around a pipe and an offset in it to accommodate a cable tray. There was a platform suspended off the column's side and braced by pieces of angle iron. Above the platform a single light bulb hung on a wire. It had one of those old-fashioned turn switches on it. A note on the drawing pointed to the light and said "To illuminate platform." Another notation near the platform identified its purpose: "To change light bulb"!

Field modifications are a fact of life simply because people do change their minds. But more often than not, something got missed in the design stage that needs fixing. The disgruntled designer may complain about the changes he has to make in the office, but sometimes that is nothing compared to what needs doing by the field forces, who in turn will most certainly blame the design office.

In the heat of construction, necessary but minor modifications are made without notification to the design office. A stairway is moved or deleted, a pipe is routed to make room for maintenance, or a chute does not fit properly and is redesigned. Sketches are made and filed, but nobody gets around to having the original drawings brought up to date. This can be hell for designers six months or six years down the road when a major expansion or modification takes place. In order to minimize such problems, it pays to have a set of "as-built" drawings made.

An efficient way to do this is to bring in a team from the design office for a period of, say, six weeks to do the job. They should work on a complete set of transparent reproductions of the original design rather than on the original drawings, because then there is a very definite separation of field changes and normal design changes. This divides the responsibility between office and site, so the work should be supervised by the construction manager.

Sometimes "as-built" incorporates changes that are made as a result of commissioning and start-up activities. If this is the case, then "as-built" should be defined as when the plant reaches production capacity or is officially accepted by you. If a long program is anticipated to reach full capacity, you may elect to pick an arbitrary cut-off date.

CLOSING REPORT

If you have ever bought a new house, you have noted that your contractor will fix a certain number of its deficiencies (if he is honest). But if you keep coming at him with minor problems, at some point in time you will notice that he just doesn't show up any more. The minor problems that you have—a loose shingle, a plaster crack, a dripping faucet, a door that warped — keep growing. You begin to fix them yourself rather than wait for the contractor, but always there is some little thing. The project ended for the contractor a long time ago, and it did for you, too, only you didn't know it. You merely slipped into an operating and maintenance phase without realizing it. It would have been a lot better if the contractor had simply sent you a letter stating that his part was over. This may have raised your ire, however, so he just let his part of the project die quietly.

Similarly, in a relatively major project there is a tendency for the project to go on and on ad infinitum. While contractors may love the extra work, your overall budget is slipping. You should, therefore, very firmly declare that the project is finished. If you want to create a new project of plant modifications or upgrading, that is fine.

The submission of a closing report is rare because most operators are preoccupied with day-to-day problems. But it is the only way to formalize the success or failure of a project. Such reports are valuable for the information they provide for future projects. Even a successful project has mistakes that should not be repeated, but more importantly, if a project is a success, why is it a success? What was done right that can be a model for the future? Unsuccessful projects will generate their own closing reports, but even then, operators sometimes are too preoccupied with problems to take the time to prepare one.

Closing reports should focus on four main points:

- Cost vs. budget
- Schedule
- Production capacity
- Product quality

Each of these requires elaboration. The following is a suggested table of contents:

1.0   Introduction
2.0   Criteria
3.0   Plant description
4.0   Cost vs. budget
5.0   Schedule
6.0   Engineering
7.0   Procurement
8.0   Construction
9.0   Start up
10.0   Production capacity
11.0   Product quality
12.0   As-built drawings
13.0   Engineers and contractors
14.0   Summary
15.0   Conclusion and recommendations
16.0   Photographs and appendix

The *introduction* should be a general preamble, describing what the report is for and what it includes.

Criteria should include the scope of work as well as the parameters defined at the outset of the project.

Plant description should summarize the final plant: building sizes and process.

Cost vs. budget should include an overall cost compared to the original budget and the latest budget update. Then a summary of final costs should be given by area and separated into engineering, home office, and construction.

Schedule should be compared to original milestone expectations together with budget updates. Discussions of major discrepancies should be included.

Engineering quality, costs, and productivity should be discussed, along with any major deviations in schedule.

Procurement highlights and problems should be outlined.

Construction quality, costs, productivity, and setbacks or major problems should be reported.

Start-up success and any problems encountered should be mentioned.

Production capacity and product quality may require a detailed discussion.

As-built drawings, if any, should be listed and their cost included.

Engineers and contractors should be evaluated for their performance. The report should state whether the same groups would be considered for future work.

The *summary* should give an overview of the events that took place on the project.

The *conclusion* involves a discussion of how satisfactory the project has been plus *recommendations* for this project and future projects. You might mention, for example, that you foresee a plant expansion in the future.

Photographs of key areas are always interesting. Also, an *appendix* of illustrations or drawings can be added to illuminate the report.

And finally, a list of the people who wrote the report, the key personnel who managed the project, and the people to whom the report has been circulated should be included.

CHAPTER

20

LOOKING AHEAD

Project management has a bright future, whether or not there is a vigorous, expanding economy. Old plants will always need replacing or updating. One thing that is certain is that things will continue to change: There is nothing more constant than change. And management of those changes is our challenge. As owners, we will continue to look for efficient ways to make capital expenditures. As consultants, we will look for the means to hone our skills.

WHO IS GOING TO RUN PROJECTS?

Many owners of plants which have gone through major, aggressive expansions over the years have tried doing their own project management, engineering, and construction and had mixed success. But there is always the problem of internal productivity — a subject that is easier to deal with by using consultants because you can fire them — and the problem of what to do with project personnel during nonexpansion periods. And there is always the threat that an empire will be created which must be dismantled at considerable expense. Other owners have suffered at the hands of less than competent consultants and have resolved to take a more active role in working with them. One extreme case is in the nuclear industry in Canada, where the client — a government agency — seems to have an antibody for every body in the consultants' groups.

In operating companies it has generally become known that an internal transfer to running projects can be the kiss of death in terms of one's future. Almost always, personnel leaving one sector of the company for projects find themselves in a redundant role in the eyes of the company once the projects are completed.

Since there is sometimes a reluctance to put a consulting group in complete charge, many companies hire an experienced project manager on contract to act as the owner's representative on their project. He uses a team gleaned from the owner's personnel. This can only be as successful as the owner's ability to choose the right man, but the idea is gaining popularity and is expected to become more common in the future.

COMPUTERIZATION

There is an almost fanatic pressure to computerize project management, and this is likely to increase in the future, though very often for the wrong reasons. Computers are fashionable, for one thing, but they also can make life easier for the project team. Two questions have to be asked: What is the payback? Who is going to operate the computers?

In terms of payback, the consulting groups stand the best chance of cashing in on the benefits of the computer, simply because projects are their day-to-day bread and butter. Owners, on the other hand, must evaluate the short-term gains even if they envision projects until doomsday, because economies — as we have been rudely shocked to learn — have a habit of doing an about face when you least expect it. What benefit is there in having CAD (Computer-Assisted Design) capability if there are no foreseeable projects? And worse, if you have purchased the hardware, your projects get shelved, and there you are, watching helplessly as your equipment becomes obsolete before your very eyes. There are signs that consulting companies may lease their equipment and software directly to owners, which could solve this dilemma for those who insist on doing their own work.

Who is going to operate the computers? We have a generation of hands-on engineers coming out of the schools who use computers with a high degree of competence. Most problems in the past have been caused by the fact that there was always an interface (a computer operator) between the engineer and the printout. In the future, we can expect our engineers to be these young whiz kids, and they will become our seasoned project managers. Despite the hype, however, we are not there yet. And furthermore, we can expect a change-over that is not entirely painless.

DEUNIONIZATION

Construction is experiencing and will continue to experience a trend towards nonunion labor. We must take care that the confrontations of the past do not develop again. The only way to do that is to make sure that labor is represented in the decision-making process.

LUMP-SUM PROJECTS

As far as overall projects are concerned, there is a trend developing towards the turnkey, lump-sum project being bid by consortiums of consultants, suppliers, and contractors. In order to reduce risks to plant operators, front-end work will be done by consulting groups to the point of accurate scope definition on a cost-plus basis, followed by open bidding for the total work.

MODULAR DESIGNS FOR MEGAPROJECTS

Megaprojects will not go away, despite their dismal record. However, the methods of approach are being rethought. Economies of scale are being reanalyzed on the basis of the trade offs between cost savings and the high cost of bailout when things go wrong. The problem of what to do with the large body of personnel that has been trained for megaprojects is being examined.

A second look is being taken at modularizing large projects. If, for example, a project such as a major tar sands plant is broken into mining, extraction, and refinery, and each of these sections is built simultaneously, then the complex logistics of personnel and materials dictate a long schedule, on the order of seven or eight years. Yet each section by itself could be done in a shorter period, say three years. By building them all at once, the managers, engineers, project management services, and construction personnel must serve a seven or eight year period instead, and a meal is made of the job. As a result of this concept, some megaprojects are now being considered in sequenced packages, that is, modular, expandable designs are being made.

Certainly, there will be plenty of projects, both large and small, in the future, and we should look forward to participating in them.

SUGGESTED READINGS

*THE COMPETITIVE EDGE
by The Associated General Contractors of America,
1957 E Street N.W.
Washington, D.C. 20006.

COST ENGINEER'S NOTEBOOK
by American Association of Cost Engineers,
308 Mononganela Building,
Morgantown, West Virginia 26505.

MANAGING CAPITAL EXPENDITURES FOR CONSTRUCTION PROJECTS
by Kenneth M. Guthrie.
Craftsman Book Company of America.

MORE CONSTRUCTION FOR THE MONEY
by The Business Roundtable,
200 Park Avenue,
New York, N.Y. 10166.

THE MYTHICAL MAN-MONTH — ESSAYS ON SOFTWARE ENGINEERING
by Frederick P. Brooks, Jr.
Addison-Wesley, 1974, Reading, Mass.

PROJECT AND COST ENGINEER'S HANDBOOK
by F. C. Jelen.
American Association of Cost Engineers, 1979,
Morgantown, West Virginia.

*THE SECRETARY
by Canadian Construction Documents Committee,
85 Albert Street
Ottawa, Ontario
K1P 6A4

ZEN AND CREATIVE MANAGEMENT
by Albert Low.
Playboy Pbks., 1982,
New York, N.Y.

---

*Contract document standards are available through each of these organizations.

INDEX